D0458168

PAX TECHNICA

PHILIP N.
HOWARD
PAX TECHNICA
HOW THE
INTERNET OF
THINGS MAY
SET US FREE OR
LOCK US UP

Yale
UNIVERSITY
PRESS
NEW HAVEN
& LONDON

Published with assistance from the Mary Cady Tew Memorial Fund.

Yale University Press books may be purchased in quantity for educational, business, or promotional use. For information, please e-mail sales.press@yale .edu (U.S. office) or sales@yaleup.co.uk (U.K. office).

Set in Joanna type by Newgen North America, Austin, Texas.
Printed in the United States of America.

ISBN 978-0-300-19947-5 (cloth : alk. paper)

Catalogue records for this book are available from the Library of Congress and the British Library.

This paper meets the requirements of ANSI/NISO Z39.48-1992 (Permanence of Paper).

10 9 8 7 6 5 4 3 2 1

For Gina Neff, who makes things possible and worthwhile.

CONTENTS

PREFACE

In the next few years we will be immersed in a world of connected devices. This book is about the political impact of having everyone and everything connected through digital networks. The "internet of things" consists of human-made objects with small power supplies, embedded sensors, and addresses on the internet. Most of these networked devices are everyday items that are sending and receiving data about their conditions and our behavior. Unlike mobile phones and computers, devices on these networks are not designed for deliberate social interaction, content creation, or cultural consumption. The bulk of these networked devices simply communicate with other devices: coffeemakers, car parts, clothes, and a plethora of other products. This will not be an internet you experience through a browser. Indeed, as the technology develops, many of us will be barely aware that so many objects around us have power, are sensing, and are sending and receiving data.

One industry analyst estimates that the internet of things will have an installed base of twenty-six billion devices by 2020, only a billion of which will be personal computers, tablets, and smartphones. An industry consulting firm estimates thirty billion connected devices. One of the main manufacturers of networking equipment estimates fifty billion devices and objects. In the next five years more than a thousand networked "nanosats"—relatively small satellites that operate in formation and have low transition power—will be launched into space. Drone production, whether for the military or hobbyists, is difficult to track. But government security services have them, and activists and humanitarian organizations have them, too. A report from the OECD on the internet of things estimates that a family of four will go from having an average of ten devices connected to the internet now to twenty-five in 2017 and fifty by 2022. Every one of those will have sensors and a radio that can broadcast information about the time, the device's location, its status, and how it has been used.[1]

Industry estimates like this are often bullish. But it is safe to say that by 2020 there will be around eight billion people on the planet, and three or four times as many connected devices. Engineers expect so many of these connected devices that they have reconfigured the addressing system to allow for 2 to the 128th power addresses—enough for each atom on the face of the earth to have 100 internet addresses.[2] The internet of things is developing now because we've figured out how to give everything we produce an address, we have enough bandwidth to allow device-to-device communications, and we have the capacity to store all the data those exchanges create. But why write a book, now, about the politics of the next internet?

Many of us are not happy with the internet we have now and are eager to find more ways of protecting individual privacy, sharing data, and bringing access to everyone. The internet of things, with embedded sensors and extensive device networks, will solve some problems but exacerbate others. Many of the design choices for this next internet are being made now, and our experience over the past twenty-five years is that it is almost impossible to use public policy to guide technology development after the technology has rolled out to consumers. And there are clues—there is *evidence*—about how the political internet has developed that can help us anticipate the problems and think proactively about how to steer the massive engineering project that is the internet of things.

For example, the latest smartphones, watches, and wearable technologies reveal how immersive and pervasive the internet of things will be. Cell phones have the ability to take one location point per second, but don't do so because their power supply is limited. If you give an application on your phone permission to use location information, it will send information to a server at the rate the developer chooses. If you use a crowd-sourcing application for traffic data, your phone is sending data about your commute. If you use an application to keep track of your jogging, your phone is generating geotagged data about your movements relative to other people. Every time you take a picture, check in with your favorite social networking application, or track your health, data is sent from your phone to a cell phone tower or router and over a vast network of digital switches.

More important for political life, the data flows through many different kinds of organizations: the companies that maintain your digital networks, the startups that build the apps, the third-party advertising agencies that have licensed access from your service providers and the startups. The platform developers and major social media organizations, such as Google, Facebook, and Microsoft, also have data access—at several points in the flow of information. The National Security Agency and perhaps other governments or other uninvited organizations can tap in.

The current objective for geolocation engineers is to design chips that require so little power that they can be left on all day. This would mean being able to generate one location point per second, all day long. As the price of making small, relatively simple chips declines, more chips can be put into devices other than your cell phone.

The internet of things will be the next, immense, physical layer of networked devices. We experience the internet through a few kinds of devices and the browsers they support. But the internet of things will be defined by communication between devices more than between people. It will be a different kind of internet: larger, more pervasive, and ubiquitous. What will be the political impact of such connectivity? What can we learn from the past twenty-five years about politics and technology that might help us anticipate the challenges and opportunities ahead?

For now, there's little research, experience, or public conversation on how the internet of things should be developed and organized. Scenarios are easy to imagine, especially since we know how media ownership issues have played out. For example, Google bought Nest, a home-automation company, for $3.8 billion in 2014.[3] Nest makes household thermometers that connect over the internet. Having one of these sensors in your home might help manage your heating needs. But it might also allow Google to know when you are home, which room you are in, and when you leave your home. Rather than imagine troubling scenarios, I want to develop the basic premises of how digital media have affected our political lives so far. Then I want to use those premises to understand the likely consequences of rolling out the new infrastructure of an internet of things. If we don't have a public conversation about the politics of the internet of things, we risk being trapped by decisions made for us.

I wrote this book because I believe that while the internet has been used in many places to creatively open up some societies, it has been used to close down too many societies with censorship and surveillance. My goal is to focus not on the internet you are familiar with but on the one engineers, computer scientists, and technology designers are working toward. The underlying assumption of this book is that while the rapid diffusion of new information technologies may disrupt political life in the short term, there should eventually emerge some noticeable patterns of behavior among political actors, some consistent trends in political life, and some conservatively safe premises about how global power is going to work in the years ahead.

In other words, new technologies like mobile phones and the internet may seem disruptive. But our political scientists, policy makers, and pundits have developed routines and habits for dealing with the technology-induced chaos of current events. The usual trope is to say that the world is falling apart, politics will never be the same, and disruption will be the norm in the years ahead. But perpetual disruption probably isn't a rule, and the sooner we can see beyond these apparent moments of chaos, the better.

I have been investigating the political impact of new technologies for two decades, and my fieldwork has taken me around the world. I've interviewed tech-savvy activists, privacy gurus, and government censors. My first investigations took me to Chiapas to meet with the Zapatistas and learn about their internet strategies in 1994. More recently, I traveled to Russia to meet with Putin's nationalist youth organizers to learn about their internet strategy. I was one of the few researchers calling attention to the crisis of popular discontent over digital media in Tunisia and Egypt in 2010. After the Arab Spring, I served as an election observer in Tunisia. I have done interviews with technology entrepreneurs, government regulators, and privacy advocates in Azerbaijan, China, Hungary, Singapore, Tajikistan, Tanzania, the United States, and Venezuela.

Although I have specialized in social science and am interested in science and technology studies, I hope the audience for this book will be broad. Because of my training, I have taken care with citations. Also, I've made much more use of public Creative Commons material than many academics do. Wikipedia and Project Gutenberg have been particularly useful for this, and I won't apologize for using them. With the growing quality of open-source manuscripts and crowd-sourced definitions, I'm comfortable recommending these kinds of references to readers who want to check my facts. Using such online material means readers will have access to updated sources even after this book has been published. Where needed, I provide more traditional sources. Some trends may change, with implications for my argument. But I don't think they will change much.

As the internet of personal computers, tablets, and mobile phones meets the internet of things—everyday objects made "smart" via sensors and silicon—what will it all mean for governments and citizens? For some, the internet is making the world a more uncertain and dangerous place. The diffusion of the internet, mobile phones, and a host of new networked devices has left many of us feeling cynical and unfulfilled politically. I challenge this and make the opposite argument: the world is coming together; politics will never be what it once was, but a new order will emerge, and there is more to be said than simply that the path ahead is scary. My goal in this book is not prediction but prescience. We'll know in a few years whether some of the trends I identify play out as the latest evidence suggests they will.

The internet of things could be the most effective mass surveillance infrastructure we've ever built. It is also a final chance to purposefully integrate new devices into institutional arrangements we might all like. Active civic engagement with the rollout of the internet of things is the last best chance for an open society.

INTRODUCTION

The "internet of things" is the rapidly growing network of everyday objects that have been equipped with sensors, small power supplies, and internet addresses. We are used to an internet of computers, mobile phones, gaming consoles, and other kinds of consumer electronics. But a whole host of other products are now being connected in digital networks, including cars, refrigerators, and thermostats. And many industries are looking for ways to put cheap wireless chips into their products in ways that will impress shoppers.

This will transform the internet. It will go from being an information network we deliberately use through a few dedicated devices we think of as "media" into a pervasive, yet hidden, network of many kinds of devices. Wired and wireless devices will be everywhere, embedded in a range of everyday objects, and therefore less visible.

We are entering a period of global political life I call the *pax technica*. I've coined this pseudo-Latin term to capture my broad theory that the rapidly increasing diffusion of device networks is going to bring about a special kind of stability in global politics, revealing a pact between big technology firms and government, and introducing a new world order. As with the Pax Romana, the Pax Brittania, and the Pax Americana, the pax technica is not about peace. Instead, it is about the stability and predictability of political machinations that comes from having such extensively networked devices. The pax technica is a political, economic, and cultural arrangement of institutions and networked devices in which government and industry are tightly bound in mutual defense pacts, design collaborations, standards setting, and data mining.

Technological innovation has historically given some countries the upper hand in global affairs. But over the past decade, technological control and information access have consistently become the key factors explaining political outcomes, and no particular country seems to have the upper hand. The smart, stupid, or surreptitious use of digital media by political actors consistently has the biggest impact on who gets what they want. Such new world orders have been given the label of *pax*—an epoch of predictable stability based on known rules and expectations. The internet of things is establishing a new pax.

Popular uprisings against long-standing dictators have rocked the Arab world. Antiestablishment movements in the West—the Tea Party in the United States, the Pirate Party in the European Union, the Occupy movement globally—have organized protests and captivated voters in unexpected ways. Around the world some regimes are more precarious, yet others seem as stable as ever. The Western internet, constituted by billions of mobile phones, computers, and other networked devices, has formed the largest information infrastructure ever. But this great device network has rivals and attackers. Battles between rival network infrastructure from China and competing norms of internet use from Russia, Saudi Arabia, and other authoritarian political cultures will dominate political life in the years ahead.

In this book, I argue that nation-states, polities, and governments need to be thought of as sociotechnical systems, not representative systems. We are used to thinking of politics as a process by which a few people represent the interests of many people, either through some democratic process or by fiat. But the internet of things is increasingly reporting on our actual behavior, generating politically valuable data, and representing our habits, tastes, and beliefs. Political communication is no longer simply constituted by citizens and politicians. Political communication systems are coordinated by network devices that citizens and politicians use with varying degrees of sophistication. We are launching such a system now, in the internet of things. Ours is the pax technica. In this new era, it may make less sense to speak of unambiguous categories of democracy and dictatorship. Instead it may be most revealing to characterize a government on the basis of its policies and practices regarding network devices and information infrastructure.

Governments are technical systems that tie their work to territories bounded by borders and claim a monopoly on certain kinds of technical expertise, information, and military power. Their character is defined both by the people who work in government service and by the material resources and devices that are built to support their administrative practices. On the whole, democratically elected governments are comparatively open technical systems, and authoritarian regimes are relatively closed technical systems.

Indeed, a spectrum of regimes from "open" to "closed" may capture more of the important nuances in what makes a contemporary government than a spectrum that gauges levels of "democracy" and "authoritarianism." The surveillance scandals triggered by Edward Snowden and Chelsea Manning and the censorship tactics exposed by the OpenNet Initiative complicate many governments' claims to being democratic.

Thinking in terms of democracy and authoritarianism does not make sense in a world where authoritarian governments use digital media to measure and respond to public opinion in positive ways. A growing number of regimes permit no public displays of dissent or high-level elections but do build new ways of interacting with citizens, encourage involvement in public policy, permit digital activism on particular issues such as pollution and corruption, and allow local elections for minor offices. The democracies in the pax technica maintain their stability by using digital media for surprising levels of social control: political and corporate data mining, digital censorship, and online surveillance are some of the activities we all wish democracies wouldn't do.

A closed government is one in which the norms, rules, and patterns of behavior for government personnel are hidden from public view and difficult for outsiders to access. Democracies and dictatorships alike protect government agencies from domestic or international challenges. A relatively small number of people— usually from the networks employed by the government—get to determine the other items on the agenda. A significant number of government processes are dedicated to protecting or furthering that government agenda. An open government shouldn't have these features.

Civic groups, journalists, and the public need to be ready. Governments are going to have less and less ability to govern the internet of things. Corporate actors, and bad actors, could have enormous power. Right now government and corporate priorities dominate. Civic groups are getting better at expressing their concerns with how the internet is developing. But they need to be ready for the internet of things.

Why is our imagination about the future of global affairs so rife with images of technological fixes to social problems? It is usually best to describe technological innovation in terms of evolution, rather than revolution, and the same caution should be used to describe the changing nature of international affairs.

Global politics has evolved significantly over the past twenty-five years, often because of technology diffusion. Too many pundits downplayed these changes because they want to see proof of the direct causal connections between rapid technology diffusion and instant political outcomes. I suspect they will be waiting for a while. But I am certain that they are missing the important aspects of institutional innovations for which there is a clear trail of evidence. Indeed, ignoring the impact of digital media on the organization of global politics is a dangerous strategy. Waiting is not a good mode of civic engagement—especially if the internet of things materializes as projected.

So what should we do? Internationally, we must actively engage in the process of setting global technology standards, encourage as much openness and interoperability as possible, and relax overrestrictive copyright regulations. And now we need to concern ourselves with the ownership structure of mobile phone companies, especially in countries where the government directly owns those companies. Media pundits used to rail against cross-ownership of newspapers, radio stations, or television stations. When infrastructure companies also produce content, net neutrality—the idea that all data on the internet should be treated equally—is at risk.

In the first chapter, I define the internet of things. I talk about the ways in which our political lives are already being affected by this rapidly growing, barely noticeable network of devices. In the second chapter, I analyze the important developments in technology and politics during what I call the internet interregnum: the period after the collapse of the Soviet Union in which our internet grew from a network of computers into a network of mobile phones.

The next internet, the internet of things, is going to allow us to draw even more nuanced maps of the most meaningful social networks. In the third chapter I map out some of the new relationships among people, data, and the internet of things. Chapter four moves from observations and examples to the conservative generalizations we can make about technology diffusion and political communication. In this chapter I offer five basic premises about how we use the internet in politics, and it is important because these premises render the likely consequences of the internet of things. In the fifth chapter I explore five reasonable political consequences of the emerging world order, this pax technica. What are the political consequences of an internet of things? The pax technica is not a guarantee of peace so much as a sociotechnical structure for political life, and in the sixth chapter I identify the major challenges to the stability of the evolving internet of things. The final chapter concludes with some reflections on how important it is for citizens and civil society actors to fight for their place in this new world order.

Every new technology seems to challenge our democratic values in some way. We sense threats to our privacy, and see greater potential for social control and political manipulation. But the internet of things will also provide greater opportunities for challenging power and building the institutions we want to build. What kind of new world order will emerge when everyone, and everything, is connected?

PAX TECHNICA

1 EMPIRE OF CONNECTED THINGS

Behind every empire is a new technology. Empires build information infrastructures that connect distant towns with major financial centers. Network infrastructure allows innovations from any one part of the realm to quickly benefit everyone else. Entrepreneurs have an easier time figuring out supply and demand. Military leaders have an easier time defending the empire, and political leaders have more information about public needs.

The Romans had such an empire, because they built the roads and aqueducts that provided their empire structure and stability. The British also had such an empire, because they had a network of fortifications and a superior navy to manage their global trade routes.

In this chapter I map out the expanding infrastructure of networked devices. I explore this domain with some hard data—a kind of census—on the size of the empire of connected things. Then I discuss some of the ways these networks of connected devices get used in political ways. When governments fail to protect us, and are unable to even warn the public of danger, we use digital media to build new systems of early warning. When governments are working well, they can overreach and use digital media to censor and surveil their citizens. Increasingly, we find that device networks are pressed into service for political and military objectives.

Carna Surveils the Realm

New information technologies have transformed world politics, and not always in good ways. Even trying to understand how technology connects us reveals the strengths and weaknesses of the internet we have built for ourselves. To understand what the internet is becoming, let's start with a basic question—how big is it?

Recently, a creative programmer decided that it might be an interesting exercise to count all the devices that were connected to the internet. We still don't know who did this internet census, but for now let's just call her Amanda. Completing an internet census was an intellectual and engineering challenge. Most important, she wanted to do it without causing trouble—she wanted to ping devices without interfering with them or slowing down the internet. So she built a "bot" and created a "botnet."

The word "botnet" comes from combining "robot" with "network." A botnet is a collection of programs that communicate across multiple devices to perform some task. The tasks can be simple and annoying, like generating spam. The tasks can be aggressive and malicious, like choking off internet exchange points, promoting political messages, or launching denial-of-service attacks. Some of these programs simply amuse their creators; others support criminal enterprises.

In playing around, Amanda discovered a surprising number of unprotected devices connected to the global internet. She realized that the only way of doing a complete census was to build a botnet that would enlist all the unprotected devices in the service of the census project. So she wrote a chunk of code that would both count devices and replicate itself so that its copies could help count devices. When she activated the bot, the botnet spread out and found 1.3 billion addresses in use by devices around the world.[1]

Amanda called her script the Carna Bot after the Roman goddess of health and vitality. For her, the exercise was about taking basic measurements of the health of the internet. Her bot worked brilliantly, reporting on many different kinds of devices, from webcams and consumer routers to printers and door-security systems. Amanda decided to remain anonymous but published her findings as a public service. Even though she had noble goals, she exposed two dark secrets about how the internet works.

First, she revealed that knowing the default passwords for pieces of key equipment could give someone access to hundreds of thousands of consumer devices and tens of thousands of industrial devices around the world, from gaming platforms to industrial-control systems. So as the world's security experts debate the impact of the latest sophisticated hacking attempts from China or the encryption possibilities of quantum computers, just knowing factory passwords means someone can access any device once it leaves the factory and is connected to the internet.

Second and more concerning, the bot discovered other bots. Carna wasn't the only unauthorized bot checking for open ports on devices around the globe. Amanda's bot was written as a public service for an exploratory project, and it built a botnet to do the census. She found several competing botnets, and an enormous, sleeping, network of bots called Aidra, which had compromised as many as thirty thousand devices. Aidra had the power to hijack not just computers but gas meters, refrigerators, microwaves, car-management systems, and some mobile phones. The bots could attack any network infrastructure for a client with a denial-of-service attack. Amanda had her Carna Bot perform the public service of temporarily disabling any Aidra bots it found.

The next time someone reboots those infected devices, the bots will be ready to start commandeering them. The botnet that Amanda exposed could be very destructive if it is ever used, and some might even see her as a threat because she was fooling around with the world's device networks. Still, in exposing these dark secrets, Amanda revealed a lot about what our internet is becoming.

What's in a Pax?

The Pax Britannica was a period of history, between Napoleon's defeat and World War I, during which the British Empire managed global affairs. London was the center of power, the British navy controlled the most important sea-trading routes, and relatively efficient bureaucracies put the world's resources and people into the Empire's service. Several aspects of the Pax Britannica may actually describe our future as much as that moment of our past.

The British were strong because their network infrastructure gave them unparalleled levels of political, economic, and cultural control. The Pax Britannica was hardly a period of universal peace—it was a period of stability more than peace. There were nasty, violent brushfire wars throughout the British Empire as poor communities resisted the oppression of colonial masters. Rival kings, separatist movements, nationalist causes, and radical socialists (and anarchists, for that matter) constantly challenged the authority of the British crown. When the Pax Britannica finally waned in the middle of the twentieth century, these conflicts between allies and challengers had lasted more than a century and cost millions of lives. The sun never set on the Pax Britannica, but it cost a lot to maintain that network of colonies.

For a long century British control of global exchange yielded great profits and political stability. Alliances among Europe's royal families managed global empires and worked out diplomatic routines, enabled faster communications between power centers, and created a tacit understanding of who controlled what.

The stability of the Pax Britannica made a few people much richer than everyone else. Economic productivity improved overall, and advances in medicine and public institutions resulted in longer life spans and more democratic engagement in the Empire's colonies than in previous centuries. Certainly development was uneven, resulting in glaring inequalities on the basis of gender, race, and faith. Economic wealth was concentrated in northern urban centers. The colonies funneled riches back only to these centers. In the end, the Pax Britannica produced a set of global institutions that still serve northern wealthy countries better than they serve the global south.

A pax evolved because government and industry interests were closely aligned. The people at the center of all this were a curious mix of technocrats, princes, and business elites. The organized faiths of the Catholic and Anglican churches also provided some social structure. These helped to form community bonds that connected core and periphery, and gave those from the core who traveled to the periphery an existential sense of mission. But the social structure provided by the church was not unique to this period of history. What was unique was the rise of a new organizational form, the "firm."[2] Moreover, these new firms were tightly coupled with the state, such that the East India Company, for example, was able to marshal the resources of the British navy for the company's global operations, and British fortifications provided homes for the Hudson's Bay Company.[3]

A pax indicates a moment of agreement between government and industry about a shared project and way of seeing the world. The key here is that the shared project involved infrastructure. It wasn't simply an agreement between governments and businesses to help each other. The collective project involved each putting information infrastructure to work for the other's needs, building it and guarding the project in mutually supportive ways, investing in innovative new technologies, and applying existing technologies, all in such a way that would shore up the power of each while allowing others to benefit.

A few people in the center of the pax made decisions about the development of opportunities at the periphery. Most communities in the rest of the world had limited control over their own development. The British government made key decisions. It defined the borders. It decided which countries would get which technologies and resources. The British military broke down resistance to international trade. This created what historians call "path-dependent development." For most of the world, the needs of the center limited economic growth at the periphery. Indeed, many of the communities that were simply producing staples—such as minerals and food—regressed in quality of life and fell apart socially.

These defining points hold for other periods that historians have described with similar labels, such as the Pax Romana or Pax Americana. While these periods of political stability were marked by border skirmishes and outbreaks of violence between local power brokers, there were few large-scale wars. Governments and firms worked together to develop new communication technologies. There were widespread benefits to the new information infrastructures, and the elites who participated in this pact benefited most. The social forces behind rival empires and breakaway republics—each seeking to build or restore its own competing network of power—were a constant threat.

And it is safe to say that the Pax Americana is over. Historians have used this label to refer to the dominance of the United States in international affairs since the end of World War II. In important ways this period of stability (more than peace) occurred because the United States managed to dominate global industry, finance, and culture. Some would say that the collapse of the Berlin Wall marked the peak of the Pax Americana, and that the internet is just an extension of America's ability to wire up economic, political, and cultural life in other countries for its own benefit.

Device networks now provide more of that structure than cultural exports. Today, governments and the technology industry have been closely collaborating on foreign policy. Indeed, in important ways, technology policy has become foreign policy. In recent years, the U.S. State Department and Silicon Valley have found more and more creative ways to work together. They fund and develop research projects together. They exchange personnel. And executives from the State Department and Microsoft, Google, Apple, and other big technology players often share the stage at public events. Increasingly, they subscribe to the theory that technology diffusion and democratic values reinforce each other and spread together. An open, global internet is good for business and good for democracy. But as I'll argue, the United States has lost control of this digital project in important ways. The United States is no longer the primary source of innovation in digital networks and the most important builder of information infrastructure. The internet no longer just "speaks English," and the Pax Americana is probably over.

The Demographics of Diffusion

More than ever, technology and technical expertise mean political power. Political clout now comes from ownership or regulation of mobile-phone networks, and control over the broadcast spectrum. The technology trends are well known but still impressive. By 2015 more than a billion people are on Facebook, and every day a half million more join. YouTube has 500 million unique visitors every month who view 95 billion videos. Every minute, users upload more than 3,000 images to Flickr, to say nothing of the other kinds of multimedia content that visitors upload to other variations of blogs, feeds, and websites. Twitter handles 500 million tweets per day, and 12 new accounts appear every second. When new social-media technologies are developed, they can attract millions of users in a blink of an eye. It took Google+ less than three weeks to attract 10 million users.

Yet it is not a particular tool or application that has created these unique circumstances of history. Altogether, the suite of digital technologies allows such levels of interactivity, creativity, and access. Moreover, usage patterns vary around the world. Facebook is only slowly making inroads into Russia and Brazil. The Chinese government has built rival platforms for almost all of the interesting digital media technologies developed in the West, so that its security services can use digital media for social control.

In 2000, only about 10 percent of the world's population was online. By 2015, more than half the world's population has internet access, two billion people have smartphones, and almost everyone on the planet has a mobile phone or easy access to mobile technology through family and friends.[4] Three of every five new internet users now live in a politically fragile country, but people have used digital media to strengthen family and friendship ties, build political identities of their own, and make other kinds of social groups more cohesive.[5] Digital media have changed the way people use their networks and have allowed them to be political actors when they want to be engaged. They use the technologies to connect to one another, and to share stories.

For decades, the greatest flow of digital content was between London and New York. That's changed now, too. The majority of traffic once flowed through the undersea trunk cables connecting North America to Europe.[6] However, the network has evolved quickly as more and more devices have become connected. In the past few years, Asian cities have been demanding more bandwidth than cities in the West, and the majority of the world's internet users live in those Asian cities. February 2013 was an important month for the new world order, because it was probably the last time that the West dominated the use of global bandwidth. If you live in the West, this was the month you lost this centrality. If you live in China, this was the month your region became the dominant network center. If you live in other parts of the world, it was the day in which the center of your economic, cultural, and political universe shifted.

Before February 2013, the bandwidth used across the trans-atlantic cables that connect the United States with Europe averaged just under twenty terabits per second. Most of that traffic was between the United States and the United Kingdom. Relative to other parts of the world, these two countries had the most internet users, the most internet servers, and the fastest networks. Much of that data involved market-exchange data between financial centers.

A month later, the bandwidth being used by the cables connecting Asian countries averaged more than twenty terabits per second. The most important center in global networks—measured just in terms of bandwidth—shifted from the West to

Asia. The bits themselves don't care how they travel: their job is simply to flow between digital switches. And the routes they take can only be estimated with probability models. There is no consciousness-raising singularity here. But at the end of March 2013, more traffic flowed between China, Korea, and Japan than flowed between the United States, the United Kingdom, and Europe.

A significant amount of digital traffic flows through the cables at the bottom of the Pacific Ocean. However, it's not undersea cables that carry the most traffic, it's overland cables. The fastest-growing region for internet traffic is within Asia—between China's largest cities and between China's cities and other cities in Asia. Indeed, one of the more pressing infrastructure problems in China is the need for faster internet connections from Shanghai, Hong Kong, and Beijing out to the country's many large provincial capitals. These days, only a quarter of all global internet traffic flows between North America and Europe.

Information Technology and the New World Order

Information technologies are now the primary conduit for everyone's political, economic, and cultural lives. This is not true for many people, or for most people in rich countries. Since the end of the Cold War, pundits and policy makers have thrown around a variety of terms to help frame current events. Do we live in a unipolar world, where the United States gets to be dominant? Or is it better to think of our political world as a multipolar one, with many different kinds of political actors busy projecting different kinds of power? The collapses of the Berlin

Wall in 1989 and the Soviet Union in 1991 are two events that clearly mark a transition point in global politics. But a transition to what?

Lots of people have tried to describe the new world order. Many have argued that with the fall of the Soviet Union, the new world order became one in which major political and security crises would be over economic issues rather than ideological differences. Anne-Marie Slaughter said our new world order was new because networks of public officials were connecting across international borders to more effectively solve problems.[7] Henry Kissinger said the United States would still be the most important part of the new world order. Francis Fukuyama said that history was over in that capitalism had triumphed and markets would be setting the rules. And John Ikenberry said that the new world order would be governed by liberal values and the institutions of international law that were set up after World War II.[8]

However, it's always a tough project to define a new world order, because you need to figure out what power is, who has power, and what it means to exercise power. These days, the people who design new information technology, produce content for digital media, and set internet standards have a significant amount of economic, political, and cultural clout. They can efficiently manipulate popular opinion, and they use information technology to make more effective use of labor and resources. Unfortunately, it's also a real challenge to identify the people, organizations, or countries that have this power today.

A lot of technology design happens in the United States, where Silicon Alley produces digital content that streams around the world and Silicon Valley creates the new gizmos everyone wants. The U.S. National Security Agency (NSA) has the ability to monitor global internet traffic in comprehensive ways. Yet the power to set standards for the global internet rests with a handful of opaque, quasi-governmental, global organizations like the Internet Society and the International Corporation for Assigned Names and Numbers (ICANN).[9] And plenty of technical standards with serious implications for our privacy are set by specialized engineering committees that are susceptible to corporate influence. Infrastructural challenges come from the Chinese government, which has direct control over its technology users and is exporting hardware to other countries so that its infrastructure network can grow. And there are the technology insurgents. There's a lot of digital muckraking that happens with The Pirate Bay and WikiLeaks. Each month some embarrassed government tries to deal with a major information scandal by going after hackers and whistle blowers.

There's no doubt that political communication in many countries has changed radically since the internet arrived. Blogging, tweeting, crowd sourcing, and collaborating online was once the sport of geeky narcissists. Now these activities shape the national policy agenda in most countries. By simply posting videos about abusive police or corrupt officials, people can rapidly undermine government credibility. At first, these kinds of activities were just pinpricks that had no chance of puncturing an overblown, dictatorial state. But by now, every government in the world has faced some kind of damaging scandal or been brought to heel by citizens who used their mobile phones to document the illegal or embarrassing acts of political leaders.

To understand current events, it's important to look beyond people and organizations to the technologies themselves. For the past twenty years, there's no doubt that the United States has been the central node in global networks of technology design and information flows. The most important force in global politics was the United States precisely because this was the country that was building the information technologies that everybody else wanted to use. This was a temporary state of affairs.

Pax Romana, Britannica, Americana

Maintaining power during the Pax Romana, a long period of relative stability established by Caesar Augustus in the late first century B.C., was a bit more straightforward and enduring. With a powerful army, the Romans conquered vast swaths of territory well beyond Italy. With a powerful infrastructure, the Romans ruled all of western Europe, the Near East, and North Africa. There were nasty skirmishes at the borders of Augustus's empire. Some of the borders of the Pax Romana—like Hadrian's Wall—were more symbolic frontiers of the Roman Empire than firm markers of territory. But Roman networks of economic exchange lasted for more than two hundred years (and in the East, for more than a thousand), and patterns of family alliances kept an overall power structure in place.

The Romans grew rich, and their system of roads, ports, and runners allowed for improved communications during economic and military crises. Rome made the important decisions about political life in the provinces, and the city's wealthy trading families made alliances with the institutions of government that put the legitimate military might of the state in their service. Romans worked hard to maintain their infrastructure of roads and seaports, shoring up their own power while also benefiting others.

Technology always has limits—it doesn't reach everyone and it doesn't serve everyone in the same way. And the infrastructure that allowed the Romans to have an extended period of political and economic stability was territorially bounded. It was an extensive network, but good roads and public works projects ended where the Visigoths began. Rome was not and could not be everywhere. Having power during the Pax Romana meant having some control over the nodes in the Empire's networks, and as a city, Rome was the confluence of these networks of power. Similarly, in the British Empire, London served as the node, and the big corporate players all managed their affairs from the capital. The fashions, designs, and innovations of London radiated outward to the colonial seats of British power.

Cultural exports from the United States were an important part of the Pax Americana: Hollywood movies, television programs, music, and advertising techniques had a significant impact on the values of viewers, listeners, and consumers around the world. The United States still generates lots of cultural memes, but many more now seem to come from the Global South. Of course, in cultural production there is give and take. Some of the fashions seen and heard on the streets of London, Rome, or New York at their heights of influence actually originated in the provinces, hinterlands, and peripheries of their empires. Today plenty of cultural memes that originated online have significant offline impact.

For several decades the United States has been the center of the world's internet. A great many innovative technology designs come from the United States. There, the places we associate with innovation have become iconic. California's Silicon Valley generates new applications and hardware; Boston produces valuable technology designs; and New York generates digital content. Seattle, Austin, and several other cities are also important nodes in this network of innovation. From the early 1990s through the late 2000s, the world's information had to flow through digital switches in the United States, as created in its leading centers of technology development.

These days, if you want to have a profile in modern politics, you have to be online. Corporate affairs are now largely managed over digital networks, and some corporations have their own proprietary networks. Now, new innovations in fashion and design diffuse over Instagram and Pinterest. Political power has often shifted along with technical innovation. Harold Innis and Marshall McLuhan taught us that it wasn't just new weapons that shifted political power centers.[10] New media and communication created great opportunities for cultural dominance, turning the limited rule of particular political leaders into decades of social control by generations of ruling elites.

Major concentrations in the control of public infrastructure have a label in political history—we call them empires. In the past, empires have been defined by groups of states and peoples that may be spread around the world but are governed by a relatively small network of political elites who are exceptionally good at control through communication. When new technologies support a political order that envelops many countries and the new rules last longer than any particular monarch or political leader, we call such an order a pax. What makes networked devices a unique infrastructure comparable to Roman roads or British warships? Looking at examples of what governments, battling political groups, or individuals do in these networks is illustrative. To start, let's look where government is absent and people are desperate.

The Balaceras of Monterrey

When drug wars erupt in Michoacán, Nuevo León, and Tamaulipas, it can be dangerous even to travel on the highways. Gruesome images of street battles between police and gangs have dominated the news coming out of Mexico. And once in a while there are truly horrific events when drug gangs lash out at the people publicizing their crimes—and at the public at large.

War between security services and drug gangs often means violent *balaceras*—street battles in which civilians are caught in the crossfire. Tortured bodies hang from bridges and fighters hijack vehicles at gunpoint. The streets clear for weeks at a time. Neighborhoods that once had vibrant nightlife are empty. But rather than sheltering at home alone, as people might once have, the inhabitants of these blighted places now connect and find ways of helping one another. One blogger, Monterrey's Arjan Shahani, relates how social media feeds on altruism.

> In traveling through Laredo with my family recently I felt a bit more protected every time a notification came in [to my mobile phone] from a traveler a few miles in front of me noting that there

was no danger ahead. With no hidden agenda and nothing to earn from it [Twitter] users I have never met such as @Gabsinelli @labellayelibro and @lacandonosa kept me and my family safe during the trip. All I can do is publicly thank them for it. Following suit, I repaid the favor and used the appropriate hashtags to provide similar information for the benefit of those traveling behind me. To all of those who selflessly participate in this chain of collaboration and communication for the better good, thank you.[11]

Maybe it is better to say that altruism feeds on social media. Even this fleeting example illustrates how online reciprocity and trust spring up in difficult settings. People gave in different ways: some contributed to a map of confirmed car hijackings.[12] It is important to examine when and how such altruism erupts over networks, and what its limits can be.

Monterrey has always been a tough town, but as soon as the city had a handful of Twitter users, the events of the drug war quickly trended.

There are reports of blasts on Venustiano Carranza Avenue #Shooting #RiskMty #MtyFollow

This basic tweet reported the time and location of blasts, along with the hashtags or keywords that help label events for tracing through time. And the tag #RiskMty became both a resource and a source of identity for citizens victimized by both drug lords' attacks and government responses. As always, the eyewitness accounts were most valuable, especially if they came with images or short video.

Neither the drug lords nor the government expected a network of real-time war correspondents to spring up. #Fuerza-Monterrey became the hashtag that allowed people around Mexico to encourage Monterrey's citizens to stay strong during the crisis. #TienenNombre became the hashtag for gathering the names of all the victims of the war on drugs, and it was borne out of a popular sense that the official statistics were greatly underestimating casualties.

An important part of why such hashtags sprang up is that local governments failed to provide information about the chaos. Local security services, not about to publicize their strategies, cowed news organizations. But what happened in many Mexican cities was still a kind of state failure. Or more accurately, public institutions failed to provide the information needed to help citizens feel secure. Many governments have procedures for communicating with the public during crises. (In the United States, these people are "public information officers.") When city and regional governments failed in public communication, people created their own community alerts.

These are other examples of how people create their own public alert systems. When Hurricane Sandy hit Santiago, Cuba, information didn't come from the state; it came from the country's independent (illegal) journalists. Text messages about serious damage and the loss of life circulated among people a day before state media tried to bring citizens up to date. When Hurricane Sandy hit New York City, even the capacity of a relatively strong municipal government was expanded by a crowd-sourced disaster map.[13]

Obviously, the evolution of media use in Monterrey is mirrored elsewhere in Mexico. The battles that have ravaged many parts of the country have actually helped drive up people's use of social media. Twitter traffic in Reynosa and Saltillo peaked when the gun battles in the street were most violent. In Monterrey they

peaked during the Casino Royale Massacre by drug gangs; in Ve-
racruz they peaked during a spate of kidnappings. The technology for this curation varied from place to place.

Device networks, and services like Twitter and SMS, have had a
similar impact in other global crises. During the Iraq War, people
used blogs to report on their struggles. When rumors spread
during a political crisis in Kyrgyzstan in 2010, people went online to validate stories in real time. During the Arab Spring, people in Egypt used Facebook to document the crisis day-to-day,
and to describe their fears and hopes.[14] And the crowd-sourced
Rassd News Network sprang up, a unique social–media-based
volunteer news network with more than a million followers on
Facebook and a unique SMS-based funding structure.[15]

The consequences of using social media during a security
crisis are always complex. There were reprisals against community tweeters, from both drug lords and angry police chiefs.
There were plenty of households with no Twitter users that were
unable to contribute or benefit from being online. Many community tweeters did not trust each other, or trusted only those
they had preexisting social ties to. Eventually, Twitter feeds were
clogged with advertisements and misinformation, and communities had to move on to other hashtags and develop sophistication in interpreting what was coming over their feeds.

Communities rapidly develop their own hashtags and keywords and documentary techniques. The Witness Project encourages people to document the abuses they see, and coaches
citizen journalists on how to use care with their work. In Monterrey, some tweeters were "halcons" or falcons, people who
watched Twitter traffic at the behest of the gangs, looking for

tips about police maneuvers and informants.[16] Tweeters competed to break news and correct one another.

Social media help people cope during civil strife. First, they follow the news and keep track of family and friends. Most of the time, people use the internet for entertainment, sports, and culture. Yet in a crisis they go online to check news stories, verify facts, and see what foreign news agencies are reporting. They ping their friends and family to make sure those living in nearby neighborhoods are safe and that those living abroad know what's going on.

Second, they pick sides. They deliberate about who is doing what to whom, and why. And they think about which side is in the right. For a few people, this means supporting the side they think should win. Some people tweet the locations of firefights as a community service; other people tweet locations of the enemy for whichever side they think should win. Some collect tips for the police, reporting the location of drug gang lookouts and drug sales points.

Third, they document. Tweeting on street violence certainly does not have the widespread impact of a punchy piece of investigative journalism from a professional journalist. But the trail of tweets, pictures on Flickr, personal blog posts, and other digital artifacts creates an archive about events that is more public, distributed, and openly contested. Crowd sourcing the production of digital maps of shootings, health needs, or criminal activity is a way of both warning the community and processing the crisis for oneself.

The internet is valuable because it provides the medium for altruism. Even a community in crisis—especially that kind of community—has altruists, and social media let those people find one another and communicate by example. Altruism and social media feed off each other. During these same drug wars, broadcast media simply reported on the violence, vacillated on who was to blame, and offered little advice on what the public should do if their neighborhood became a site of conflict. Social media, in contrast, could be creatively used by a few people to generate value for others.

Tweeting certainly didn't stop Mexico's drug war. But it helped Monterrey cope. We can't measure how important the sense of online community provided by active tweeting can be in the first few weeks of a crisis, both in providing moral support and in keeping people safe. A few citizens rise to the occasion, curating content and helping to distinguish good information from bad.

Eventually the ecosystem of the balaceras degraded, with bots, ads, and misinformation. The drug kingpins showed their frustration with the impact of social media by targeting tweeters and bloggers for especially gruesome murders. The falcons used Twitter for their purposes. Yet people migrated to other hashtags, making for a dynamic flow of content that was propelled by an eagerness to help one another.[17] The story here is not that people tweeted and saved the world—the story is about altruism and survival strategies. Will the internet of things be a conduit for altruism?

The Internet Is Also a Surveillance State

While social media has helped communities connect during a crisis, it also provides big governments with a powerful surveillance tool. Edward Snowden gave us two lessons about how information technologies are used in the name of national security. The first is that the intelligence services don't have a magic decryption key that unlocks everybody's secrets. They have the political, economic, and social ties, or at least leverage, to get secrets out of the businesses that own and operate the infrastructure. The second is that most intelligence-technology services are contracted out to private firms. It's this security industry that builds much of the world's information infrastructure and analyzes much of the world's data.

In fact, Snowden, as a civilian employee of the National Security Agency, was one such contractor. Since 9/11, the technology and security industries of the United States have grown significantly. According to the *Washington Post*, in 2013 almost a million people had top-secret security clearances to gather information about national threats.[18] *Post* reporters counted some 1,271 government organizations and 1,931 private companies working on programs related to counterterrorism, homeland security, and intelligence in about ten thousand locations across the United States. They found that in Washington, D.C., and the surrounding area, thirty-three building complexes for top-secret intelligence work are under construction or have been built since September 2001.

All this adds up to the equivalent of three Pentagons or twenty-two U.S. Capitol buildings—about seventeen million square feet of space. Redundant information systems make several security and intelligence agencies do the same work. For example, fifty-one federal organizations and military commands, operating in fifteen U.S. cities, track the finance and information networks of terrorist groups. The volume of information these agencies study is impressive, and so is the volume of information they produce. The best estimate is that some fifty thousand intelligence reports are produced each year—a volume so large that most are ignored.[19]

Snowden also taught us that "metadata" is valuable. The content of your email and browser cache can help analysts make inferences about your behavior. The secretive Foreign Intelligence Surveillance Act (FISA) courts ordered Verizon to hand over the metadata on call duration, direction, and location of subscribers. Such a trove of metadata can be used to map out who you know and where you are, and to find patterns in your daily routines.

The Snowden affair is an example of how the weakest link in national security is social, not technological. Obviously, he was the one who leaked immense amounts of information about secretive programs. He also revealed the deep interconnections between the U.S. security and technology industries. As the cryptographer Bruce Scheier says, "Whatever the NSA has up its top-secret sleeves, the mathematics of cryptography will still be the most secure part of any encryption system."[20] He argues that the National Security Agency must spend most of its efforts on the real vulnerabilities in any digital media: poorly designed cryptographic products, software bugs, and bad passwords.

Ultimately, the best way for a national security agency to crack any security system is to get the company that designed the system to collaborate by leaking all or part of the keys and giving access to the computers and networks in question. So the NSA did not always break into the secure systems of other governments or the technology firms that built the global information infrastructure. Snowden's leaks have revealed that the NSA has used everything from polite requests to legal pressure to ensure collaboration.

And when social engineering or software bugs don't allow access, government agencies can actually buy data from internet service providers. Recent estimates are tough to come by, but in 2006 the General Accounting Office revealed that some $30 million had been spent in contractual arrangements with information resellers. So not only does the federal government fail to regulate data miners very closely, it buys products and services from these same data miners. It's safe to assume that much more is spent now, in what is largely an unregulated government procurement process.

But for a conspiracy theorist, anyone who has read *Wired*, or a historian of revolution, it is not surprising that governments and industry collaborate to sustain each other's power and to send each other business. What is surprising is the degree of deliberation behind the process, especially between the U.S. government and the country's technology firms, and their success in rolling out an internet that could be so easily commandeered in the service of the country's national-security aims.

Working out the terms of the collaboration is not always easy. Some internet companies respond quickly when the government serves real-time "electronic surveillance" orders rather than take the risk that the National Security Agency will install a hardline tap in their server rooms.[21] We know of only a few occasions when major technology firms have resisted government queries, or the gag orders that usually come with such queries.[22] A few businesses, particularly those that try to serve customers with ultrasecure systems, simply shut down. Lavabit shut down rather than comply, and Silent Circle suspended its operations because its managers thought subpoenas, warrants, security letters, and gag orders were likely to come.[23]

Before Snowden, there's little doubt that technology companies responded quickly when queried by national security agencies—especially in the United States. More recently, several firms have been issuing public statements on the number and kinds of government requests they get, and openly discussing government access to customer records in newspapers and with watchdog groups. In 2013, Facebook reported more than twenty-six thousand requests from governments around the world. That year, Facebook responded to 43 percent of all government requests for data, and 80 percent of the requests coming from the U.S. government. In 2014, Google produced some data for 65 percent of all government requests and 84 percent of those coming from law enforcement agencies in the United States. That year, Twitter complied, in some way, with 52 percent of global requests and 72 percent of U.S. requests.[24]

Whether or not you think Snowden was a traitor, his impact has been to teach us all how big and powerful our high-tech surveillance state has become. He also taught us that government and technology firms were more than cooperating and colluding: they were conspiring to find ways to work more closely with government and to advance each other's interests.

The internet is not a cloud; it is made up of buildings with expensive air-conditioning systems that help keep racks of equipment cool. Undersea trunk cables connect continents. Private businesses own and maintain those cables. Those businesses have lobbyists who try to acquire bigger and bigger contracts from governments. A simple way to describe this relationship is to say that politicians made investments in surveillance firms, and surveillance firms made investments in politicians.

Of course, public infrastructure projects always involve political decisions about how much to commit from the public purse, and whom to employ to make dreams about the public good a reality. Surveillance firms started making big investments in lobbying for more surveillance. They successfully tied—in the public's mind—the idea of making people safe with the idea of watching for trouble. After the terrorist attacks of September 11, 2001, being safe meant tolerating surveillance.

Snowden's revelations broke that link. People are now much more aware of the degree to which the government can access their personal records and communications. Public vitriol on this issue may not be enough to drive significant change in how the U.S. government conducts itself, and U.S. allies do not consistently express outrage about being surveillance targets. Ultimately, Snowden's contribution has been to help teach the public about technology and surveillance. For political actors that do not have the infrastructure to respond with equal measures of surveillance against the United States, there's another way to act.

The Wars Only Bots Will Fight

If you can't comprehensively surveil or censor the internet, the next best strategy is to write automated scripts that clog traffic, promulgate your message, and overwhelm the infrastructure of your enemies. One unique feature of the emerging political order is that it is being built, from the ground up, for surveillance and information warfare. Another is that it has new kinds of soldiers, defenders, and combatants.

The ongoing civil war in Syria has cost hundreds of thousands of lives. The great majority of these are victims of President Bashar al-Assad's troops and security services. After the Arab Spring arrived in Syria, it looked as if the forces loyal to the Ba'ath government would stay in control. Speedy military responses to activist organizations, torturing opposition leaders and their families, and then the use of chemical weapons seemed to give Assad the strategic advantage. But even with these brutal ground tactics, he was unable to quell the uprising in his country. And by 2013 he was losing the battle for public opinion, domestically and abroad. Even China and Russia, backers that supplied arms and prevented consensus in the U.N. Security Council about what to do, succumbed to political pressure to join the consensus that al-Assad had to go.

What's unusual about the crisis is that it might be the first civil war to have been fought by both human combatants and bots. Public protest against the rule of Assad, whose family had been in charge since 1971, began in the southern Syrian city of Daraa in March 2011. Barely a month later, Twitter bots were detected, trying to spin the news coming out of a country in crisis.

In digital networks, bots behave like human writers and editors. People program them to push particular messages around, or to respond a certain way when they detect another message. They can move quickly, they can generate immense amounts of content, and they can choke off a conversation in seconds. From very early on, people in Syria and around the world relied on Twitter to keep track of what was going on. Journalists, politicians, and the interested public used the hashtag #Syria to follow the protest developments and the deadly government crackdown. The countering bot strategy, crafted by security services loyal to the government, had several components.

First, security services created a whole host of new Twitter accounts. Syria watchers called these users "eggs" because users' pictures remained the default image of an egg. No real person had bothered to upload an image and personalize the account. Regular users suspected these profiles were bots because most people do put some kind of image up when they create their profiles. Leaving the default image in can be the marker of an automated process, since bots don't care what they look like. These eggs followed the human users who were exchanging information on Syrian events. The eggs generated lots of nasty messages for anyone who used keywords that signaled sympathy with activists.

Eggs swore at anyone who voiced affinity for the plight of protesters, and pushed pro-regime ideas and content that had nothing to do with the crisis. Eggs provided links to Syrian soap opera TV shows, lines of Syrian poetry, and sports scores from Syrian soccer clubs to drown out any conversation about the crisis. One account, @LovelySyria, simply provided tourist information. Because of the speed at which they work, the pro-regime bots started to choke off the #Syria hashtag, making it less and less useful for getting news and information from the ground.

A little investigation reveals that the bots originated in Bahrain, from a company called Eghna Development and Support.[25] This is one of a growing coterie of businesses offering "political campaign solutions" in countries around the world. In the West, such companies consult with political leaders seeking office and lobby groups who want some piece of legislation passed or blocked. In authoritarian countries, "political consulting" can mean working for dictators who need to launder their images or control the news spin on brutal repression. Eghna's website touts the @LovelySyria bot as one of its most successful creations because it built a community of people who supposedly just admire the beauty of the Syrian countryside; the company has denied directly working for the Syrian government.[26] But @LovelySyria has few followers and not much of an online community presence. With two tweets a minute, Twitter itself decided to stop @LovelySyria from devaluing a socially important hashtag.

Of course, automated scripts are not the only source of computational propaganda. The Citizen Lab and Telecommix found that Syrian opposition networks had been infected by a malware version of a censorship circumvention tool called Freegate.[27] So instead of being protected from surveillance, opposition groups were exposed. And a social media campaign by a duplicitous opposition cleric was cataloguing his supporters for the government.

One estimate holds that 75 percent of all Twitter traffic is generated by the most active users—about 5 percent of all Twitter accounts.[28] In terms of Twitter accounts, one-third of those active users are believed to be machine bots each tweeting more than 150 times a day. Because some bots generate fewer than 150 tweets a day, the actual number of bot-held accounts is probably higher. In terms of Twitter messages, as many as one-quarter of all tweets sent in an average day may come from bot accounts.

Most of these crafty bots generate inane commentary and try to sell stuff, but some are given political tasks. For example, pro-Chinese bots have clogged Twitter conversations about the conflict in Tibet.[29] In Mexico's recent presidential election, the political parties played with campaign bots on Twitter.[30] An aspiring British parliamentarian turned to bots to appear popular on social media during his campaign.[31] Furthermore, the Chinese, Iranian, Russian, and Venezuelan governments employ their own social-media experts and pay small amounts of money to large numbers of people to generate pro-government messages.[32]

Even democracies, as Snowden revealed, have groups like the United Kingdom's Joint Threat Intelligence Group.[33] These shadowy organizations are also charged with manipulating public opinion over social media with automated scripts. Sometimes Western governments are unabashed about using social media for political manipulation. For example, USAID tried to seed a "Cuban Twitter" that would gain lots of followers through sports and entertainment coverage, and then release political messages by using bots.[34]

Keeping track of bots is hard work. We know they are out there. The Storm Botnet of 2007 infected as many as fifty million computers.[35] It was reportedly powerful enough to force entire countries off the internet, and there was speculation that its builders—as yet unidentified—were getting ready to sell access to some of the bots' abilities. A year later, the Kraken Botnet grew to more than 400,000 bots and generated nine billion spam messages a day, often from the computers of Fortune 500 companies.[36]

These days, the safest assumption is that there are lots of bots that we don't know about. We can only hope that the best of them are dedicated to spam rather than political interference. Storm and Kraken are now several years old, but they are still going, and they mark a turning point in the recent history of bots: many believe that they were designed to target the security companies and spam-listing services that the rest of us rely on to tell us which bots need to be blocked.

These days, it is not uncommon for bots to attack the organizations that try to keep the internet healthy and working well. Unfortunately, most of the world's webmasters depend on one organization—Spamhaus in the Netherlands—for accurate records to keep Viagra ads and appeals from wealthy Nigerian widows out of our email and social media feeds.[37] Increasingly, Spamhaus itself is the target of attack because disabling its watch list would allow pernicious bots to flood the global internet.[38]

Only a few bots operate like Amanda's Carna Bot—the one that gave us the first real census of the internet of things. Other bots have been designed as weapons of war, and as weapons in propaganda wars. They are tough to spot and tougher to shut down; sometimes they are put up for sale. These weapons of mass digital disruption operate beyond public policy oversight. The International Telecommunications Union (ITU) may attempt to generate modern communications policy, the ICANN may help to adjudicate internet addresses, and the Internet Governance Forum (IGF) may help internet stakeholders talk through technology standards.[39] But real power over technology rests with businesses, the people who build nefarious or noble bots, and the political leaders who deploy armies of bots for their projects.

The Political Empire of Connected Things

On an average day, the United States and all of its allies face automated attacks by vengeful enemies. Some of the attackers are foreign governments, criminal organizations, or politically motivated hackers. The targets are a wide range of government agencies and public infrastructures. Increasingly, the targets are civil-society organizations, because it is often civic groups in democracies that draw the most attention to injustices in less democratic regimes.

Many different kinds of political actors are aggressively using social media more and more not just to reach their followers but to involve them in the organization.[40] This was one of the fundamental political innovations of Obama's campaign for the U.S. presidency in 2008, and an innovation adapted by well-funded civic groups around the world.[41] The Lebanese army has an app for uploading geotagged images of suspicious cars in the neighborhoods of Beirut it controls.[42] The next version will allow users to download images of "wanted" political figures. So it's not simply about putting out propaganda. It's about organizational incorporation.

In today's physical battlefield, information technologies are already key weapons and primary targets. Smashing your opponent's computers is not just an antipropaganda strategy, and tracking people through their mobile phones is not just a passive surveillance technique. Increasingly, the modern battlefield is not even a physical territory. And it's not just the front page of the newspaper either. The modern battlefield involves millions of individual instructions designed to hobble an enemy's computers through cyberwar, long-distance strikes through drones, and coordinated battles that only bots can respond to.

The reason that digital technologies are now crucial for the management of conflict and competition is that they respond quickly. Brain scientists find that it takes 650 milliseconds for a chess grandmaster to realize that her king has been put in check after a move. Any attack faster than that and bots have the strategic advantage—they spot the move and calibrate the response in even less time. The most advanced bots battle it out in financial markets, where fractions of a second can mean millions in profit or loss. In competitive trading they have the advantage of being responsive. Slight changes in information, however, can result in massive, cascading mistakes.[43]

Whether it is competition in stock markets or over social-media feeds, we are finding that bots are increasingly dominating digital networks. These networks provide us with our news and information, culture, and more. This doesn't mean that they always capture our attention. They can certainly compete for it, clog our networks, stifle economic productivity and block useful exchanges of information. We are living in the last empire because a majority of the cultural, political, and economic life of most people is managed over digital media. Control of that information infrastructure—whether it's by technology firms, dictators, or bots—means enormous power.

Billions of devices connect to the internet, and enormous botnets of mysterious and malicious code are ready to wreak havoc. The center of global information infrastructure is shifting away from the United States, with the largest volumes of traffic and the majority of technology users now in Asia. Past empires were defined by the cultures and communities that built innovation technologies and used them for economic and political gain. But what is emerging now is a political order constituted by the relationships between devices as much as the relationships between people. Obviously, technology networks grow along with social networks, but that means our digital devices increasingly contain our political, economic, and cultural lives. These devices provide some significant capacities, but also some troubling constraints on our political future.

Technology diffusion has had many different kinds of political consequences in countries around the world. But one global consequence is an evolving pax technica. The primary cause of the pax has been the diffusion of the internet of things—diverse devices that many of us no longer even notice as being part of the internet. The beginning of this story is demographic—almost everyone is online. The internet is the conduit for modern political culture and conflict, and almost everyone is affected by this new global network. The next part of the story involves collusion, conspiracy, and crisis.

Infrastructure has a significant impact on the constraints and capacities of political actors. So understanding the new world order means digging into the recent history of how networked devices have proliferated. Understanding how the internet of things will impact our political lives requires a look at its technological substrate—the political internet that we've built for ourselves over the past twenty-five years.

2 INTERNET INTERREGNUM

The world seems more chaotic than ever. Hackers take down important websites while large businesses and foreign governments spend big money on cybersecurity and cyberespionage. Our own governments use the internet to track our activities. Technology companies conduct mood-manipulation experiments. The Russian mafia buys our credit card records so they can figure out where to steal the best cars. Chinese hackers take secrets from governments and intellectual property from businesses. Drones help catch terrorists, but also violate our sense of privacy. The internet was supposed to change politics forever, but every new app seems to expose us to new risks.

But we've actually just come through the era of real uncertainties—a kind of interregnum. It was a twenty-five-year stretch between the political order of the Cold War and the beginning of something new. In 1991 a group of hard-line Communist leaders tested Mikhail Gorbachev's reforms in the Soviet Union. Dedicated citizens wouldn't give up their cause and kept up their acts of civil disobedience. Boris Yeltsin made an impassioned plea from atop a tank in front of Russia's parliament buildings, and the hard-liners lost. Yet that was also the year that Tim Berners-Lee published the first text on a webpage and demonstrated how large amounts of content could be made widely available over digital networks. Within only a few years, idealistic new social movements like the Zapatistas were using the internet to advertise their struggle and build international audiences.

Twenty-five years after the Zapatista Rebellion, many popular uprisings for democracy and homespun activist campaigns were marshaling social media for political change. Political elites were using digital media too, and the first bots went to war for their political masters. The Zapatistas had organized offline, found the internet, and used it effectively for propaganda. Movements like the Arab Spring were being born digitally.[1] They were organized online, and projected power in the streets of Tunis and Cairo. Osama bin Laden, the most wanted global terrorist in recent memory, was caught and killed because the internet of things betrayed him. A new world order, of people and devices, had emerged from the uncertainty of technological transition.

Discovering the UglyGorilla

This new world order includes the UglyGorilla—he is a hacker, not a great ape. He has been at work since at least 2004. He's a high-profile member of China's cyberarmy, and the only reason we know he exists is because he has made mistakes. In February 2013 the security firm Mandiant released an extensive study of China's hackers and their impact on governments and businesses in the West.[2] UglyGorilla, also known as Wang Dong, featured prominently in the report because investigators managed to get access to the passwords used for his various accounts. They were able to trace his activities over an extended period.

Along with his teammates in People's Liberation Army Unit 61398, UglyGorilla shares responsibility for multiple attacks on businesses, news organizations, and governments. He has infected devices around the world with different kinds of malware. In May 2014, the Federal Bureau of Investigation (FBI) in the United States issued a wanted poster for UglyGorilla. He and several colleagues were indicted, the first time the FBI had charged the employees of other governments with crimes.

As a student, UglyGorilla had taken classes at China's National Defense University, where he asked a professor in the Department of Military Technology and Equipment whether China had cybertroops.[3] Just being a credible programmer and making this query opened a new career path, and within a few years Ugly-Gorilla's moniker was appearing in several different kinds of malicious code.

He is not the only one to follow such a career path. Hundreds of students at other universities across China ended up in the same unit after making queries to their cadre leaders and professors. As a team they have hijacked thousands of devices around the world, but Mandiant was able to track the hackers' digital footprints back to a run-down office building in a district of Shanghai. They've stolen medical records, blueprints for new computer chips, and strategy documents.[4] By 2015 they had hit the infrastructure of more than one hundred firms and dozens of government offices. They have hacked the New York Times for Chinese Party officials. They hit civil-society groups too, especially human-rights groups in the West that work on the plight of Tibetans and the treatment of other minorities in China.[5] They have gone into the computers of defense contractors in the United States, and into the communications infrastructure of the public and private utilities that run our power and water supplies.

Assessing the harm of these attacks, the value of lost intellectual property and damaged equipment, and the costs of increased security is tough.[6] We do know that the largest organized team of government-sponsored hackers can be found in China. Chinese firms are also victims: hackers working for public-private partnerships and Chinese startups routinely go after one another's intellectual property.[7]

China is not the only country to blame for escalating cyberespionage. Other governments help their country's businesses with industrial espionage, and those firms assist governments on security issues—as I have stressed, this is the basic deal behind the pax technica. We also know that Western firms and governments collaborate on industrial espionage.[8] Hackers from eastern Europe, many with connections to Russian crime syndicates, have breached Apple, Facebook, and Twitter. North Korea sends its elite hackers, specially trained at its military school, Mirim College, to attack South Korean and U.S. infrastructure on national holidays.

The list of state-sponsored viruses is growing. One attack crippled the world's most valuable company, the $10 trillion Saudi oil firm Aramco. Hackers wiped out data on three-quarters of the company's computers.[9] The attack was probably launched by Iran, and it came on a carefully chosen day when the impact would be severe. Stuxnet, the virus that crippled Iran's uranium enrichment centrifuges, was probably developed by the United States and Israel.[10] The same team that produced Stuxnet probably also produced the viruses Flame and Gauss, all of which have some shared code.[11] These more recent viruses have basic data-mining goals, and Gauss seems to be targeting Lebanese banks.

China is only one of several countries that have a full-time, professional cohort of hackers who aggressively attack information infrastructure in other countries and steal intellectual property. Along with the United States, Israel, and China, the government payrolls of Russia, Bulgaria, Romania, and Ukraine are believed to include hackers.

We can learn from UglyGorilla. First and foremost he has taught us that "national security" must now include the ability to respond to cyberattack, and that the initiators and targets of such attacks are not just governments. Cyberattack involves finding and exploiting vulnerable device networks by entering, copying, exporting, or changing the data within them. The distinction between corporate espionage and state espionage is no longer so meaningful, and Western technology firms and media outlets are among the most prized targets.

Second, he's shown us how businesses and governments in the West are increasingly bound up in a kind of mutual defense pact. Organizations within the pax technica are ever more mutually dependent for cybersecurity. Governments and private firms are forced to share ever more information about the kinds of attacks they undergo and the security standards they maintain. State, business, and civil-society actors in the world's democracies are deeply interdependent regarding cybersecurity.

For better or worse, cybersoldiers such as UglyGorilla often have several employers during their career, sometimes going to work for one Chinese company against another. So the good thing about collaboration within the pax technica is that governments, firms, and civic groups rarely attack one another. They may surveil each other, but they often share security knowledge.

The discovery of some background on UglyGorilla doesn't make it more likely that he will be caught, though the FBI does want him caught.[12] With a host of programmers boasting such valuable skill sets, it is more likely that he'll generate new pseudonyms, change workstations, and move offices. His network is here to stay.

UglyGorilla has been throwing his weight around the international arena. This shadowy figure is not the only new kind of political actor with influence. Another distinctive feature of this internet interregnum has been the blossoming of unusual kinds of social movements employing the creative use of digital media. Malicious state-sponsored hackers are not the only disruptive force in modern international affairs.

Devices of Hope

For a modern social movement to succeed—since movements compete with other movements—network devices are needed both for organizational logistics and for reaching an audience. Mobile phone–wielding activists used to inspire a lot of hope. It seems like only yesterday that an aspiring insurgent with some basic consumer-grade electronics and a decent data plan could bring any urban center to a standstill, or toss out even the most recalcitrant dictator. Now it seems as if some movements for popular democracy have lost their technological advantage.

These days, mobile phones, drones, hacktivists, and cyber-attacks seem simply to add to the chaos. Many strongmen and authoritarian governments have quickly climbed the technological learning curve and put their digital devices to work as tools of social control.[13] Regimes in Iran, Bahrain, and Syria use Facebook to expose opposition networks and to entrap activists. China, Russia, and Saudi Arabia make big investments in surveillance infrastructure, with national, internal internet structures built from the ground up as tools for cultural management. Recently, the U.S. Department of Defense announced that cyberterrorism had replaced other forms of terrorism as its primary security concern. Drones are only the latest technology to challenge our domestic policies on airspace, privacy, and access to consumer electronics, to say nothing of challenging our warfare ethics.[14] Technology firms are earning a bad reputation for the ways they experiment on and manipulate public opinion.[15]

Anonymous, a group of online activists, has become a force in global affairs.[16] It targets whomever its members want, sometimes with impact. In 2014 Anonymous took down soccer's World Cup website to protest poverty, corruption, and police brutality in Brazil.[17] It has exposed corruption, gone after child pornography websites, and embarrassed cults. Its operations are not always successful, sometimes doing damage more than raising awareness. When overwhelmed, governments have begun to address Anonymous as an equal in negotiation. The government of the Philippines has tried to engage with the group by making concessions and involving it in national cyberstrategy.[18]

Internet pundits have added to the chaos of international politics. Julian Assange's online WikiLeaks project exposed diplomatic correspondence and upset many delicately balanced relationships among states and between power brokers. Both Assange and Edward Snowden decided that democracies were the least likely to provide them with just treatment as whistle blowers. The Russians gave Assange an online talk show and have sheltered Snowden. Moreover, many kinds of authoritarian regimes like Russia now employ their own social media gurus to engage with the public. Having more information and communication technologies hasn't made international affairs more transparent, honest, or democratic. If anything, global politics seems even more convoluted and complex with the arrival of the internet.

Rather than bringing clarity to our understanding of where global politics is headed, technology pundits have made complex trends even more confusing. Malcolm Gladwell declared that political conversations don't count for much unless they are face to face, something recently deposed dictators, recently elected politicians, and active citizens would dispute with personal experience. Media pundit Evgeny Morozov has excoriated the U.S. State Department for spending taxpayer dollars on technology initiatives, even though the number of groups grateful for the affordances of new technology tools grows year by year.[19] Gladwell's essential claim is that social-movement organizing can employ different technologies. Social media aren't particularly important because collective action happens with face-to-face contact, and social-movement logistics can be handled with a paper and pencil. I argue the opposite—that a host of political actions cannot be undertaken with paper and pencil.

The Demographics of Diffusion

While hacktivists with specialized tech skills—like the Anonymous collective—can have a dramatic impact during a crisis, the public now uses device networks for their consumption of political news and information in very straightforward ways. In the United States, the Pew Internet and American Life Project reports ever rising numbers of people promoting civic issues, contributing political commentary, and sharing news articles over networks of family and friends.[20] By 2015, YouTube had well over a billion regular users a day from around the world.[21] Every contemporary political crisis is digitally mediated in some way, and not always in a good way. Especially in times of crisis and uncertainty, people seek more sources of information. In Syria, mobile-phone and internet subscriptions continued to grow throughout the civil war, confirmation that in times of crisis people want more information.[22] When the president of Turkey, Recep Erdoğan, shut down his country's Twitter networks in 2014, he drove up public interest in learning how to use Twitter to get around state interference. When the president of Egypt, Hosni Mubarak, shut down his country's internet during the Arab Spring, he drove more Egyptians into the streets of Cairo and ultimately lost his job.

The global spread of cellphone towers and wifi access points has been rapid. By 2015 there were almost 100,000 towers, a thousand cellular network companies, and well over a million wifi access points.[23] When closed countries do open up, information-deprived communities clamor for internet access. Now that civil-society actors in Myanmar have more room to maneuver, one of the hottest countrywide social-movement campaigns has been for cheaper mobile-phone rates. People want to connect.

On top of this, there are several big demographic trends that explain why the internet of things is emerging as it is. First, device networks are spreading quickly. While our planet now holds just over seven billion people, it is home to more than ten million mobile digital devices.[24] The number of stationary—yet connected—devices is estimated at forty million. This is both an immense network of opportunity and a significant infrastructure for abuse. It's also a growing network. The world will have more wired, intelligent devices in the coming years. The network is already made up of more devices than people, and the devices continue to share information even when we aren't personally using them.

Second, more and more people are online. By 2020 everyone will effectively be online. Most people will have direct internet access through mobile phones and the internet, but everyone will be immersed in a world of devices that are constantly connected to the internet. Already people who don't have direct access are tracked and monitored through government and corporate databases. Their economic, political, and cultural opportunities are still shaped by digital media. This means that for the first time in history, virtually everybody can connect to virtually everyone else. Most countries have upward of 80 percent internet penetration. And this is only the internet of mobile phones and computers. The internet of things will keep everyone networked constantly.

Third, most internet users—and eventually most people—will soon be "digital natives." Digital natives are the people born since the turn of the century in countries where digital media forms a ubiquitous part of social life. By 2010, the majority of all internet users were digital natives. Until recently, internet users were largely a kind of "digital immigrant" population that had started going online when they arrived at their universities or had become connected through their work. This is changing. By 2010, more than 50 percent of the internet-using population consisted of people who were born into a world of pervasive internet, mobile phones, and digital media.

Alas, this population is unevenly distributed: the bulk of new technology users are young and living in Africa or Asia. Most are coming from failed, fragile states with weak economies. The norms that shaped our internet—the internet of the late 1990s and early 2000s—are going to be reshaped by the next two billion users. What will the internet look like after young people in Tehran, Nairobi, and Guangzhou reshape its content?

The Zapatistas Reboot History

In the hushed morning after New Year's Eve celebrations in 1994, hundreds of masked rebels moved through the empty streets of San Cristóbal, Chiapas. Cutting phone lines and immobilizing the local police, they wanted a new political order. Even though they hailed from the Lacandon Jungle at the southern tip of Mexico, their well-organized digital-outreach campaign put the Zapatistas into international headlines.

If the collapse of the Soviet Union marked the end of the history-making battle between capitalism and state socialism, the Zapatista uprising helped restart history by kicking off the battle over device networks. In 1995 I traveled to Chiapas, Mexico, to meet with the Zapatista insurgents. I wanted to learn about their motivations and their struggle, and to understand why they were having such an unusual impact on international politics. By the time I landed in San Cristóbal, the Zapatista Liberation Army was beginning to retreat into the jungle. Their knowledge of the forest gave them an advantage over the Mexican army. San Cristóbal was tense but quiet, and it was easy to find people who were sympathetic with the Zapatista cause.

I visited one supporter, an ecologist, at his research lab outside the city. The Mexican military had just looted his offices and destroyed his computer equipment. It was the only equipment of its kind in the region, and he was eager to talk about why his lab was a strategic target. He had used his geographic information systems (GIS) to make the only accurate maps of jungle deforestation. At the time, plotting changes using satellite images actually meant clicking a handheld pointer over the contours of a hard copy of the satellite photo—it was not an automated process. His funding came from the United Nations and was for forest-related research. In this case, plotting the growth of coffee-wealthy plantations, ranches, and logging operations was a political act. Much of the land was supposed to be collectively managed by the poor *campesinos* and *indigenas* of the region or to be under the protection of the national park system. Yet satellites could see the changes from orbit, and his lab had computed the rates of change.

The Mexican army had come for the digital maps, but the sergeant in charge didn't know what it meant for the data to be "in the computer." He thought the ecologist was hiding something, so he ordered his men to destroy all the equipment. The Zapatistas had visited him only two weeks before. They knew the value of data, and they knew how to repurpose satellite coordinates on forest cover for political impact. They had asked for the same data, knowing what they were looking for. They had brought their own diskettes for copying the data.

During that trip I found that the grievances of the Zapatistas were like those of many landless poor in Latin America. They were tired of waiting for land rights, and angry about industrial logging in the rainforest. So the Zapatistas used the internet to campaign internationally. Most of their members did not have a dial-up modem. Subcomandante Marcos, the spokesperson and nominal leader, did. When his modem broke, Marcos's speeches were smuggled out of the jungle, transcribed, and distributed by email. His eloquent, excoriating commentary activated people from around the world, and bound the Zapatistas up in a global conversation about neoliberal reform, social justice, and poverty. The Chiapas95 listservs alerted journalists around the world and kept activists engaged with compelling stories written by the insurgents themselves. The internet was used to coordinate food caravans, and to pass good stories and images to journalists. Zapatista leaders called for "pan, tierra, y liberdad," and the Zapatistas were the first social movement to go digitally viral. This global attention made it impossible for the Mexican government to put down the insurrection as violently as they had crushed rebellions from other poor farming communities in the region.

Pictures and stories flowed out of Chiapas. Most important, the compelling narratives of injustice and poverty circumvented the media blockade set up by the Mexican government. The Zapatistas didn't fully succeed as a social movement, though they now have a Facebook page. They did, however, inspire a host of other social movements to go global via the internet. Following their example, civil-society actors flourished online: fair-trade coffee, the World Social Forum, and Jubilee 2000 are only a few of the global movements that learned from the example of the

Zapatistas' digital media impact. The online propaganda success of the Zapatistas helped to establish a new set of norms for how political actors communicate and organize.

Fifteen years after talking to the Zapatistas, I traveled to Tunisia to work as an election observer. A popular uprising there had deposed the longtime dictator, Zine el-Abidine Ben Ali. For the first time, Tunisians were choosing a constituent assembly, and the excitement was palpable. Bloggers were running for higher office. There were civic projects to "video the vote." And election monitors from many different groups were capturing images from polling stations for posterity.

Both uprisings marked the start of something new. It had been fifteen years since the last "wave" of democratization. Between 1989 and 1995, many remnants of the Soviet Union and failed authoritarian regimes in other parts of the world turned themselves into variously functional electoral democracies. By 2010, roughly three in every four post-Soviet states had some democratic practices.[25] Certainly there were also large, important countries that made little effort toward democratization, strategically important states run by hereditary rulers, and other states that seemed to be slipping, sliding, or otherwise teetering on the edge of dictatorship. As a region, North Africa and the Middle East were noticeably devoid of popular democracy movements, at least until the early months of 2011.

The internet was part of the story of Tunisia's recent popular uprising. Yet it wasn't simply a new communications tool for the propaganda of democracy advocates. Many Tunisians had been disaffected for a long time, but organized opposition grew online. Digital images of the burned body of Mohamed Bouazizi circulated by mobile phone within the country and eventually across North Africa. The activists behind the Arab Spring used digital media for propaganda and organization. Their revolutionary spirit spilled across borders. Using a combination of social media and agile street tactics, they toppled multiple dictators in a surge of unrest that has been called the "fourth wave" of popular uprising for democracy.[26]

Both events are difficult to understand without considering the importance of digital media. The Zapatistas started using the internet after organizing themselves, because they found an international audience there. In contrast, the Arab Spring was born digitally.

Both movements attracted poor, disenfranchised citizens, with few land rights or job opportunities. Both groups were fighting back against authoritarian elites, who had relied on oppression and subsidies—essentially bribes—to keep their restive populations contented and dissidents marginalized. Yet the disenfranchised used the internet to catch political elites off guard. The Zapatistas did not achieve their immediate welfare and land-reform objectives, but they were successful in commanding international attention and did much to dissolve the authority of the ruling PRI party, which subsequently was voted out after ruling Mexico for more than sixty years. The leaders of the Arab Spring were successful in toppling multiple dictators, and upsetting the political status quo across an entire geopolitical region.

The stories of the Zapatistas and the Arab Spring are not about nationalist fervor inspiring political revolution. They are not about religious fundamentalism. These movements were not particularly Marxist, Maoist, or populist. They had leaders, but employed comparatively flat organizations of informal teams compared with the formal and hierarchical unions and political parties behind Václav Havel, Nelson Mandela, and Lech Wałęsa. Instead, digital photos circulated widely and kept grievances alive. Periods of political history are not easy to define. They begin and end slowly. Their features are not absolute, but are prominent and distinctive. That's how these two social movements demark the interregnum.

Despite Francis Fukuyama's claim that history was at an "end" in the early 1990s, I argue that device networks have given history a new beginning.[27] Two moments of upheaval in international affairs mark the transition. The Zapatistas used the internet to project their plight and demands well beyond the communities of Mexico's Lacandon Jungle. We've just lived through the second moment, with events in the Middle East that demonstrate how mobile phones, the internet, and social media now drive political change. A sense of frustration caused a cascade of popular uprisings during the Arab Spring.

The organizers behind the Zapatista and Arab Spring movements, twenty years apart and using very different internets, brought about change at home and upended global politics. This may not seem like an original argument because lots of people have said that the internet is revolutionizing economics, politics, and culture. People have declared that there is a revolution in how political power is organized and projected. It is important to be cautious, because as the saying goes, history is replete with turning points.

Yet carefully looking through the evidence reveals just how much our political lives have changed because of and through the diffusion of information technologies like mobile phones and the internet. We have been through a significant, technologically enabled transition, marked from the point in time when we began linking webpages to the point in time when we began linking everyday objects. The interregnum started at the end of the Cold War, as we began linking ideas and content over the internet. The transition was completed recently, when the first bot wars erupted, the Arab Spring bloomed, and Osama bin Laden was caught and killed. And now we have begun linking objects in an internet of things.

Between the Zapatista Rebellion and the Arab Spring significant features of political life changed, and a new world order, structured by device networks, emerged. Yet this transition didn't just involve protesters and insurgents. The political economy of government changed, also because of the diffusion of device networks. It used to be the amount of gold that a state stored that was the measure of its wealth, and the basis upon which that state could issue currency. Now it is information and technology. And even though it was government investment in public infrastructure that built the internet, governments no longer own and control it the way they once did.

From Gold to Bits

For hundreds of years, a government's stash of gold was the best measure of its strength and wealth. At different points in time, access to fish, timber, pelts, and other resources made a country or empire wealthy.[28] But when governments needed to borrow money to finance wars or public infrastructure, it was usually the store of gold in the crown's treasury that determined how much the government could borrow and at what rates. The result was what economic historians called a "gold standard." This single metal became the global currency because it was relatively rare and could be stored, guarded, bought, and sold. For more than a thousand years, a rich colony, country, or empire was one with lots of gold.

By 1930 this had started to change. The global economy had developed and industrialized to the point where there were many different ways of judging a state's wealth. Some countries had significant timber, coal, and fuel resources. Precious metals other than gold, such as silver, were increasingly valuable, and having good rail lines or productive factories was clearly important to a country's economic productivity. Other countries had significant labor resources, and financiers became sophisticated enough that they could evaluate a country's wealth beyond gold assets. Moreover, paper currency became a reasonably good representation of value, and economists got better at measuring the demand and supply of money, inflation, and a government's monetary and fiscal policies.

Eventually, the gold standard was not so standard anymore. Countries still maintained a supply of gold in case of a fiscal emergency, but gold was no longer the primary basis for lending money to governments. Nations had physical places where they stored gold, even after the gold standard was dropped. These storehouses of gold formed an important part of the valuation and stability of currencies well into the twentieth century. Even today, having a store of gold is one sign of sovereign wealth, though the actual amount of gold a sovereign government holds may not have much of an impact on how markets value that government's bonds or an economy's stability.

Today, other resources are regarded as having truly international, exchangeable value: technology and data. Significant amounts of value are tied up in digital networks, in remittances and electronic currencies that are not backed up by paper or gold. States once stocked up on gold to show off their stability. Now bond markets, currency speculators, and security analysts judge a government's stability by its ability to keep electrical power flowing and its devices connected to the internet. The ways of measuring a country's value have changed once again. We now have all sorts of indicators about the size of the information economy, and we often evaluate a country by how much technical innovation we find there.

The World Bank now counts patents and Ph.D.s in its country-wealth indicators. We continually worry about the supply of engineers in the economy. Bytes of traffic are a good proxy for a country's importance in the global-information infrastructure. The financial services sector expects government investment in the internet, and when countries invest in information infrastructure, many industries benefit. The perception of technical innovation, the size of the information economy, and the reach of high-tech industries are all important to the evaluation of modern economic wealth.

This new sense of valuation is what drives the startling rise of virtual currencies, mobile money, and other digital exchanges.

Such virtual currencies are designed to free money, or more abstractly "value," from the control of a particular country's central bank. The World Bank estimates that by 2020, the economy of mobile-phone money exchanges might top $5 trillion and include the two billion people who otherwise have no access to banks.[29] Some of the oldest institutions around—universities—have started accepting virtual currencies like Bitcoins for tuition.[30] It was easier for governments to hoard and guard their gold than it is data, information infrastructure, and intellectual property. States don't control public information infrastructures upon which value is exchanged.

For the first time, governments don't control the information infrastructure upon which public life is lived. They can manipulate devices, but so can many other actors.

States Don't Own It, Though They Fight Hard to Control It

Governments work hard to control device networks, even when facing pressure to relax regulations. Generally, there are four ways to grade government control over information infrastructure. The most complete control usually comes through direct ownership, and the government simply treats the national phone company and internet service provider like state assets. All information infrastructure is directly managed by the government. So the first way to evaluate state control is to see whether the national phone company, along with any broadcasters and other media outlets, have been effectively privatized. The second check is to see whether a regulator has been set up—a kind of watchdog group that tries to keep the national phone company, the

TV broadcasters, and the internet service providers honest and fair. The third check is to see whether that regulator is actually independent and free of political appointees. Ideally, the head of the regulator can be a technocrat—an engineer or a policy specialist—who is great at solving technical problems and not interested in playing politics with whatever party is in government. The fourth benchmark is the market for consumer electronics and the internet of things. This involves opening up the market so that all the latest mobile phones and wifi routers can come into the country for people to buy. If the market for the new devices with built-in power supplies, sensors, and radios is restricted, then governments have a better chance at controlling the flow of information.

In 1995, just around a quarter of all countries had privatized their phone companies or set up media-watchdog organizations. A small fraction of countries—mostly in the West—were choosing technocrats over politicians to run such agencies and letting the latest consumer electronics into the country without levying import taxes on them or imposing outright bans.[31]

By 2005, that picture had changed dramatically. Media and technology firms had successfully beat back government control over both telecommunications services and consumer electronics, and these industries had made the privatization of information infrastructure in the developing world a priority. Almost two-thirds of all countries had started privatizing their state-owned phone companies, and four-fifths had set up independent regulators. Almost half were appointing engineers over political leaders to oversee media development, and three-quarters of all countries had very liberal policies for importing consumer electronics. A few years later, not only had countries opened up the consumer market for electronics, but the technology industry itself had moved much of its production out of the countries that had built the internet in the first place. Even though most people around the world were connecting to the internet with their mobile phones, the main Western equipment makers gave up producing mobile phones.[32]

But then people started doing politics with their digital devices. Opposition movements started catching ruling elites off guard by using new communications technologies to organize huge numbers of people quickly. The very policy reforms that seemed to make some governments popular and some economies boom were allowing new political actors to act in powerful and decisive ways. After the Occupy Movement and the Arab Spring, and after a host of politicians from around the world were disgraced by the quick judgment of the internet, governments that hadn't relaxed the rules became less interested in doing so.

Lots of governments try to control the internet, and they are likely to keep on trying. They try to build surveillance systems. They try to build kill switches. They try to set the rules and regulations for developing new parts of their information infrastructures. We are rightly worried when they try. Internationally, the policy agencies that work on communication issues have faced concerned publics. Closed countries managed by political strongmen find that they can't control the internet in times of crisis, so they argue for changes in technology standards that give them a tighter grip on information flows. Fortunately, there aren't very many clear examples of regimes that have fully retained control of their country's information infrastructure.

A New Kind of New Order

The way social movements organize and grow is not the only feature of the emerging global order. Even hawkish security experts agree that security strategies have had to evolve in dramatic ways over the past twenty-five years. In November 2012 the Israel Defense Forces began its Pillar of Defense assault, and it announced the attack on Twitter first.

> The IDF has begun a widespread campaign on terror sites & operatives in the #Gaza Strip, chief among them #Hamas & Islamic Jihad targets.[33]

@IDFSpokesperson continued:

> We recommend that no Hamas operatives, whether low level or senior leaders, show their faces above ground in the days ahead.[34]

Of course Hamas responded, with both a defensive posture and a tweet.

> You Opened Hell Gates on Yourselves.[35]

Following this exchange, Israeli missiles crashed into Gaza, and Hamas mortars fell on Israel. Global alliances of opinion on Twitter could be tracked with the hashtags that also revealed allegiances: #pillarofdefense versus #gazaunderattack.

This was the first time that a Twitter battle preceded—indeed announced—a ground battle. Social media have quickly become useful tools for both insurgency and counterinsurgency campaigns. Making digital media so central to the conduct of modern warfare is only one of the distinctive features of the new order of international affairs.

In late 1996, hundreds of thousands of protesters came out on the streets of Belgrade, Serbia, to protest the official results of local elections. These protests became known as an "internet revolution" because the University of Belgrade students who organized them communicated extensively via email. They also helped journalists at the only independent radio station, B92, to put podcasts online when the government took away its broadcast license. This took place at a time when only 3 percent of the population had internet access. Two years later and oceans away, the rise of online journalism and the expansion of mobile phone access in Indonesia eroded the Suharto regime to the point of collapse. "Color revolutions" in post-Soviet countries and the Arab Spring in the Middle East then followed, each uprising involving some creative use of digital media that caught elites off guard.

Yet digital networks are for more than propaganda in political competition and conflict. Pervasive device networks helped provide much needed evidence about the location of the world's most wanted terrorist, Osama bin Laden. Vast amounts of video, photo, and text content culled from the web placed the international pariah within 125 miles of Abbottabad—a process that civilians could replicate.[36] The relative absence of mobile-phone pings to network towers drew attention to a particular compound—it was a large home that was noticeably off the grid.

Drones and satellites gave analysts the aerial view.[37] And high-risk ground observations raised more evidence.

"Big data" refers to information about many people collected over many kinds of devices, and big data helped find a major international pariah. U.S. Navy SEAL assault rifles were the proximate cause of bin Laden's death, and those military personnel took on very real risks. He was the world's leading terrorist for a decade, but the operation to get him kicked into high gear when supercomputers, state-of-the-art software, global surveillance systems, and mobile-phone networks were combined with the assets of human spy networks and elite commando units.

Today, every dictator, thug, and kingpin has experienced some kind of unwanted attention after video evidence he could not control was distributed online. Every gang, cult, and criminal network has lost members to online evidence about their bad deeds. Every activist and every revolutionary—aspiring, successful, and failed—applies a digital strategy.

Historians of revolution often argue that the most important cause of any particular revolution in a country is that country's previous revolution.[38] In other words, it is tough to identify a present circumstance that seems more causally important than whatever conditions were set up by the last big social change. Yet an important part of the new world order is that the very definition of dissidence now includes being deviant with your devices. Being a modern dissident is about having enough tech savvy to run a blog or Twitter account that tells stories different from the stories the government puts out. In some countries, having a blog can put someone on a dissident list even if the author only occasionally dabbles in political conversation or is tangentially linked to a network of dissident friends. In countries where even the opposition political parties need to be licensed, the best-organized dissidents are usually online.

Before the Arab Spring, Tunisia's and Egypt's most effective political dissidents were found online.[39] For several years, Cuba's most prominent blogger, Yoani Sánchez, has been one of the regime's most worrisome dissidents.[40] Every Free Syria Army rebel unit is equipped with a videographer. The internet is now so much more than a propaganda tool. In many countries, social media allow people to maintain their own networks of family and friends, often with little manipulation by political elites. And the data trail of Facebook likes, cell phone log files, and credit card records has proven useful for security services—the good ones and the bad ones.

But It's Not a Westphalian—or Feudal—World

Our aspirations for what information technologies might do for democracy won't go away. In 2010, Google CEO Eric Schmidt and the U.S. State Department's tech evangelist Jared Cohen argued in Foreign Affairs that global politics was in for a thorough digital disruption.[41] They urged a strong set of alliances between Western governments and global-technology firms. This, they said, was going to be the best strategy for ensuring that information infrastructure will be put to work for liberal values.

By 2020 the world will have almost eight billion people and around thirty billion devices. What will it mean to have vast amounts of political content and intelligence generated by devices and autonomous agents? Are these just more people and devices for UglyGorilla to manipulate?

To some it might seem like a classical Westphalian world. The Treaties of Westphalia, signed in 1648, ended widespread conflict in Europe and created a new political order in which rulers were sovereign over their lands and did their best not to interfere in one another's domestic affairs. Some would say that what emerged in recent years is more of a Westphalian internet, with separate sovereign internets that allow little room for outside interference in domestic technology policy.[42] In this way of looking at things, the internet has become the infrastructure for political conversations for many in wealthy countries and the medium by which young people in developing countries develop their political identities. The problem with this perspective is that these public conversations occur over private infrastructure, and the owners of our digital media can have distinctly un-public values.

Yet national governments have not simply been enlisted to advance the cause of corporate information infrastructure: such infrastructure is being put to the service of government. It makes sense that states would want to treat the internet like territory. Device networks already provide actors with the commanding heights needed to prevail, in terms of both soft and hard power.

If our current political order doesn't seem like a Westphalian collection of distinct internet states, perhaps the right metaphor can be found in feudalism.[43] After all, we are increasingly beholden to a small number of big companies for our data, connectivity, and privacy. Some of us commit to Google, others to Apple, still others to Microsoft, or sometimes our employers make that commitment for us. We commit to buying their products and services, and they commit to making new technologies available (within their information ecologies).These feudal lords try to keep our data safe.

If we want to participate in the modern information economy, or we want employment in a modern workplace, or we want to use digital media to maintain social ties with families and friends, we have to pledge allegiance to at least one of these lords. As data serfs it is difficult to pick up and move between kingdoms.

However, this perspective doesn't work because social elites are hardly using the internet to usher in a new era of feudalism. There are important civil-society actors that lead successful campaigns against the absolute dominance of technology firms. The 2012 defeat of the Stop Online Piracy Act and Protect IP Act is a good example of the clout that users can have when they are organized. An important open software movement gives us good, free tools, such as Firefox and Apache. In feudal politics the state and religious authority were one and the same thing. China's technology businesses and government agencies are sometimes indistinguishable from one another because of complex reporting structures and ownership patterns. In most democracies, technology firms and government agencies are closely aligned but organizationally distinct.

Hyperbole aside, Apple, Google, and Microsoft are not states. If sovereign authority possesses the "legitimate exercise of military power," none of these firms qualifies. The feudal state and church were deeply entwined institutions, and the technology industry and government may seem similarly bound up in agreements over data sharing, business arrangements, and collaborative standards. These days, government and technology firms collude in self-serving ways, but it is probably not a collusion legitimated by a Higher Power.

So what is the right historical parallel? Governments and tech firms definitely collaborate on foreign policy. Tech firms often get their way, but not always. Users often feel tied to their technology providers. Social networks are not bound in the same ways nations can be. Thinking of the current state of technology affairs as a Westphalian set of national internets or a highly structured feudal system misses the point: networks are essentially made up of other networks. The pax metaphor does the most to describe the network system of organizations and information technologies that enforce a de facto stability in economic, political, and cultural engagement.

A surprising stability has come from having giant technology firms and powerful nations colluding on how to sell the next generation of device networks to the rest of the world. We are entering a period of global political life that will be profoundly shaped by how political actors use the internet of things. Indeed, the internet of things will define, express, and contain this period. The capacities and constraints of political life have often been shaped by technological innovation—and vice versa. Technology and politics have an impact on each other and on how current events and future prospects should be situated in the context of the recent past. More devices come online each month, and progressively more people are connected through these digital networks. Now almost every aspect of human security depends on digital media and this internet of things.

Responsibility for creating this internet of things still rests with all of us. We use social media, and few of us are diligent about maintaining our privacy. We do our computing in the cloud and communicate through mobile phones that we know are traceable and hackable. We trust immense amounts of data to private firms and governments. Most of us have grown up with an internet that mediates political conversations between people. Increasingly, politically revealing conversations occur between devices, often long after we've finished using them. Everyday objects we aren't used to thinking of as networked devices will sense and shape the social world around them.

If the past twenty-five years has been an internet interregnum in which we developed new political habits around our communication technologies, how will this larger internet of things affect our systems of political communication? If there is going to be a great new physical layer of networked devices, we need to expand our definition of human security to appreciate the ways that this new internet of things serves us. We need to develop a public policy that ensures that this internet serves us responsibly. We need to map the new world order of the pax technica.

3 NEW MAPS FOR THE NEW WORLD

Maps are expressions of political power, both perceived and claimed. When the Romans set out to organize their expanding empire, they mapped the great lengths of roads and aqueducts that structured their social world. British cartographers provided merchants with maps of the best trading routes and equipped military officers with maps that identified the best places for fortifications. In recent years, we've started producing new kinds of digital maps that reveal new kinds of power. What new maps do we need to understand the new world order?

The usual map of the world reveals a patchwork of countries. Yet there is a surprising number of people and places that aren't really connected to the countries they are supposed to be part of. We are used to political maps that mislead us about how governments are really able to govern. Many breakaway republics, rebel zones, and anarchic territories are connected with one another in surprising ways, through dirty networks that link drug lords with rogue generals and holy thugs. Dictators are major nodes in the dirty networks of crime, money laundering, and human trafficking that are often found in the world's authoritarian regimes. Fortunately, there is a kind of demographic and digital transition taking place. The toughest dictators are getting older, while the average age of people they rule is dropping.

Social media, big data, and the internet of things are helping people bring some stability to even these anarchic places. Our crowd-sourced maps of social problems, produced by many people using many kinds of devices, help people find solutions. Increasingly the internet of things is structuring our political lives, and we can already see how people use device networks to create new maps that link civic groups with one another and with people in need. This internet can make people aware of their behaviors and relationships. It allows people to trade stories of political success and failure, and to build and maintain their own networks of family and friends. Sometimes, these networks evolve into powerful political movements. People use social media to make new maps and new movements, and to construct new institutions for themselves. Moreover, the internet of things is growing into a world where networks of dictators are calcifying, young people are using digital media to develop political identities on their own, and communities are involving information infrastructure in digital networks.

Mapping Hispaniola

Crisis is an important catalyst for the creative and civic uses of digital networks. On January 12, 2010, an earthquake devastated Haiti. The epicenter was just outside Port-au-Prince, the capital. Homes collapsed and the National Assembly building fell into a heap of rubble. Years of deforestation had left exposed soil on many hillsides, so mudslides engulfed whole neighborhoods.

The power was knocked out and telephone lines went down. The earthquake had a Richter magnitude of 7.0, and over the next few days there were some fifty-two aftershocks with devastating consequences. The headquarters of the United Nations Stabilization Mission in Haiti also collapsed, killing many, including the mission's chief.[1]

Haiti is one of the most beautiful yet impoverished countries in the world. Natural disasters take a serious toll because recovery operations are so difficult. When I was working there in the mid-1990s, the urban slums of Port-au-Prince were home to at least 100,000 people. High walls often surrounded slums like Cité Soleil and Jalouse, which held people living in extreme poverty. It was difficult to assist people in these districts, because aid groups had trouble figuring out who needed what help. Haitian Creole is a relative of French, but even if you learn the language, it can be difficult to earn the trust of the people you want to help. A few months after I left, Hurricane George devastated several large districts. The slums near the ocean flooded, causing hundreds of deaths and displacing almost 200,000 people. Recovery operations were painfully slow.

When the earthquake struck in 2010, the devastation was not only overwhelming but overwhelmingly difficult to recover from. U.N. agencies scrambled for information about what was going on. Relief workers reported from particular settlements, but didn't know what was going on in other parts of the country. Satellite images revealed the parts of the coastline that had flooded, but could not assess the health and safety needs of people on the ground. Which collapsed buildings had entrapped people? Where were people congregating for help?

Within hours of the earthquake Patrick Meier, a student in Boston, formulated a plan to help.[2] Desperate pleas for assistance were coming over the television news and his own social networks. Eager offers of help were also coming over Twitter, so why not crowd-source the problem and map out the situation? He turned his small Boston living room into a "situation room," where volunteers took queries from emergency response personnel working sixteen hundred miles away in Haiti. Fueled by coffee and working on laptops, the team wrote software for a basic crisis map that would show anyone with an internet connection both the need for and the supply of aid.

The map quickly became the centerpiece of collective action. Within a few weeks, the international community was depending on the amateur crisis map to coordinate its efforts. The application helped to rationalize efforts. It served more than an organizational function—it served an institutional function. In Haiti, the government lacked authority and organizational capacity. But for a brief period during a desperate crisis, the crowd-sourced map provided something of both.

An "organization" can be usefully defined as a social unit made up of people and material resources, like desks and telephones, while an "institution" consists of norms, rules, and patterns of behavior.[3] The map became both an organization and an institution. The digital crisis map provided an immediate logistical tool for coordinating relief workers. It helped to establish the norm that social problems could be tackled by wide-ranging, territorially dispersed volunteers interested in behaving altruistically. Social media made it possible. People put in their energy.

Device networks of mobile phones and mapping software provided structure.

When I was in Haiti in 1997, it took an extensive search to find recent maps of Port-au-Prince. The United States had made some maps twenty years earlier, but there was nothing more recent in the Haitian National Library. The Ministry of Development didn't have them, and there were none at the main university. After several weeks of searching, I found the best maps in the U.N. compound. They were under lock and key because administrators didn't want the maps to get into the hands of corrupt officials, drug smugglers, and human traffickers.

Despite the best efforts of the United Nations, even these maps inaccurately charted the capital's slums. And I was allowed to make photocopies only because the security guards liked my Canadian bona fides. Mapping social problems is certainly not a new idea. The possibilities have grown with satellite images, hobbyist drones, cloud storage, expanding device networks, and the altruism of the crowd. Since 2010, dozens of poor communities have started building new organizations and institutions through mapping and device networks. Like the slums of Haiti, the slums of Nigeria are also difficult for the government to see and serve.

Dictators and Dirty Networks

Digital networks can expose connections that political boundaries do not. Two-thirds of the world's population lives in countries without fully functional governments. Or more accurately, they live in communities not served by states with real governments. They live in refugee camps, breakaway republics, corporate-run free economic zones, gated communities, failed states, autonomous regions, rebel enclaves, or walled slums. Modern pirates dominate their fishing waters, and complex humanitarian disasters disrupt local institutions regularly. Organizing people to solve problems in such places is especially tough.

In many of these places, rogue generals, drug lords, or religious thugs who report to no one (not on this earth, anyway) lead governments. Not having a recognizable government makes it tough to collaborate with neighbors on solving shared problems. These liminal states tend to support alternative dirty networks of criminal leaders, human traffickers, and local despots.

These places have been growing in number, population, and territory. Criminal gangs, religious fundamentalists, and paramilitary groups that disrupt global peace thrive in these places. Yet these places are not always chaotic. In recent years, these locations—and there are many of them—have actually started showing that government is not the only source of governance. Information technologies have started to fill in for bad governments and become a substitute venue for public deliberation.

While population growth is making these unstable parts of the world younger, many of the world's dictators are only getting older. These days, the vast majority of the worst dictators are older than seventy or have serious health problems and no clear succession plans. Some of them control only tiny countries with few resources, but all are nodes in a network that support the growing global reach of the criminal gangs, religious fundamentalists, and paramilitary groups in their countries.

Most of them came to power before the turn of the twenty-first century. As dictators, they faced and still face a digital dilemma—a difficult choice over whether to allow their citizens access to the internet. If they do not allow internet access, the population misses out on the economic opportunities; if they allow internet access, the population gets access to news and information from countries where people live more freely. As aging dictators, they already encounter more and more open challenges to their rule from rivals and democracy advocates; internet access can only exacerbate this challenge.

What's the connection between aging dictators, malformed governments, and the internet? As the internet and mobile phones arrived in many corners of the world, people began to realize that they could make connections that their government couldn't or wouldn't make. They started communicating with their neighbors. Young people who had never known a political alternative to authoritarianism started exploring their options. People began using social media to turn political problems into opportunities. The internet and mobile phones provided a new structure for political life.

Mubarak's Choice

In the late 1990s, the global information economy was booming. Egypt was also doing well, at least compared with some of her neighbors, but was coming under pressure to modernize.

Since so much economic growth seemed to be tied to the internet, Egypt's strongman, Hosni Mubarak, faced a dilemma.

Giving the country's major businesses access to fast telecommunications services seemed like a good economic strategy. Cairo is an important financial center in the region, and many of the country's state-owned businesses would benefit from a strong communications infrastructure. Fast internet links to Europe and North America could inspire innovation and entrepreneurship at home. It would also make Mubarak look like a forward-thinking, modern leader. If the internet was good for the economy and a booming economy could help keep Mubarak in power, it made sense to invest in the internet.

At the same time, the internet also seemed like a chaotic place. There was news and information from other countries. There were reports flowing in from Egyptians abroad, not all of whom supported Mubarak's rule. Censoring the internet might cancel out the economic benefits, discouraging entrepreneurial creativity. Were the political risks worth the economic opportunities?

Mubarak and his advisers decided to go all in. They privatized the national phone company and in 1998 gave it a new name—Egypt Telecom. The company that had brought the first telegraph line to Alexandria in 1854 was now going to bring the internet to all Egyptians. It expanded its data services and invested in a mobile-phone operator.

The government tasked the Ministry of Communications and Culture with encouraging internet use around the country. It initiated a public-private partnership with Microsoft that involved lowering the cost of buying legal software to Cairo residents who could demonstrate that they were responsibly paying their utility bills. Schools and libraries got internet terminals. New fiber-optic cables were laid. The price of making cell phone calls dropped quickly. For practical reasons, Cairo was the main beneficiary of all this. Egypt has modest income inequalities, but the fastest technologies went to wealthy urban elites of Cairo and Alexandria. Most of the infrastructure investment went to those cities, and not every Egyptian could afford to get online anyway. People around the country found mobile-phone services a great new way of connecting to family and friends.

In 1998, the *Lonely Planet Guide* for Cairo listed the cheapest internet cafés and gave prices at around two U.S. dollars for an hour of access. This meant that internet use was really only for tourists and the wealthy, because the average income for people working in Cairo was around nine dollars a day. Ten years later, access was pennies an hour. And cybercafés were thriving mostly in the poorer districts of Cairo, because many in Egypt's growing middle class had access through home and work. A dilemma is a problem with two possible, equally undesirable, outcomes. Mubarak's choice to provide connectivity for citizens had political consequences.

We Are All Laila

Eman Abdelrahman carries herself with a humble and quiet demeanor. She founded, then led, a community of young women who were at the center of Egypt's popular uprising. As a young

Egyptian, she grew up only knowing Hosni Mubarak as president. She was not known as a rabble-rouser, and did not consider herself a feminist.

But in the summer of 2006, she was tired of being a second-class citizen. She and her friends were feeling less and less connected to political life in the country. Or more accurately, they were using the internet to learn about life in countries where freedom and faith coexist, and they were having political conversations through blogs and Facebook.

One day in August of that year, she was chatting with one of her friends online, and they found themselves complaining about the way men treated women in their society. As young women, they faced limited career options and sexual harassment on the streets. Without much planning, they decided to blog about their frustrations and share ideas about what could be done about their grievances. It took only a few hours to find three other female bloggers who were eager to write together. They had a group chat that night. What should they call themselves? Would they campaign for something specific, or were their goals more broadly about community building?

They all had a favorite book, Latifa el-Zayat's The Open Door.[4] In this book, the protagonist, Laila, is a young Egyptian girl who faces daily situations in which the men in her life always seem to get the upper hand. Egyptian society prevents young Laila from achieving her goals, even as she tries to have a political role in the country's nationalist movement. But she continues to believe in the importance of her role as a woman, and she continues to have faith that change will come. Abdelrahman and her friends settled on the blog name: Kolena Laila, We are all Laila.

Some pundits, including Malcolm Gladwell, argue that digital activism is inherently less productive than face-to-face mobilizing. When I met Abdelrahman a year after the Arab Spring, I asked her about the lasting impact of her virtual writers' group. She said that they had put into practice a key aspect of Egyptian political life that had been missing for a long time. They introduced—or reintroduced—the idea of writing together, publicly. Mubarak's decision to support wider internet access was especially significant for Abdelrahman and her friends. That summer of 2006 they found fifty other female bloggers willing to write short essays about their experiences as young women in Egypt. Some wrote about the daily burden of sexual harassment in the streets. Others posted opinions about female genital mutilation. Their strategy was to all write posts on the same day, referencing Laila. This way, the Arabic blogosphere would be awash with gender politics. In a coordinated effort, Abdelrahman and her friends were able to generate content, and spread that content around the Egyptian blogosphere.

Each year, the project had a bigger and bigger impact. More young women bloggers wanted to contribute, and each year they had more readers. They also had some difficulties along the way. Some male bloggers wrote in support of the group's efforts, while others disparaged it. Not all of the core writers had a steady income, and as a group they had trouble funding their collective work. They were accused of being an anti-Islamic feminist movement. They probably weren't that, but they probably were digital activists. They were an organized public effort with clear grievances who targeted authority figures and initiated campaigns using device networks.

The group managed to coordinate contributors for four years.[5] Members still savor a particular victory, in which they successfully campaigned to pressure a father in Saudi Arabia to allow his daughter to return to Egypt because she wanted to pursue academic studies. Eventually, the day of coordinated blogging became a week of collective writing and conversation. The mainstream media covered the group's activities. In its final year, Kolena Laila attracted two hundred female bloggers, from thirteen countries around the region. Despite victories like this, the group's founders still admit that cultural change is slow.

But the group's larger contribution to Egypt—and the region —was to popularize a new meme in Egyptian civic engagement: "We are all." Kolena Laila was among the first in Egypt to use digital media as a way of raising shared grievances and getting people to realize that their troubles were not theirs exclusively. Sociologists call this "cognitive liberation." People start to realize that they're not alone in wanting to improve their lives, and that they share strategies for doing so with their neighbors.[6] The internet can help cognitive liberation come more quickly.

That's because what keeps dictators and ideologues in control is their ability to make people think that they are alone in their opposition. Pluralistic ignorance—when people publicly accept an injustice but privately reject it and aren't aware that others reject it too—is a lot harder when digital networks provide constant ambient contact with others. Authoritarian governments work to make citizens believe they will be all alone if they show up for a protest.

Soon after Kolena Laila started, other internet-based organizations picked up on the phrase "We are all" for their digital campaigns. When a young man named Khaled Said was beaten up and murdered by corrupt police officers in the streets of Alexandria in the summer of 2010, a movement arose to document the attack. Photos of his brutalized body lying in the morgue circulated online. People joined the group in solidarity, realizing that their personal experiences with corrupt security services were actually a shared experience.

The number of people who supported the group grew immensely, and its organizers decided that it was time to take to the streets. They learned from the success of Eman Abdelrahman and her friends in raising public awareness. Kolena Laila had popularized the notion that you could use the internet to share your political grievances and aspirations. The group memorializing Said took to the internet. Members took their cue from the successful Laila movement, calling themselves "We are all Khaled Said." In January 2011, they brought the Arab Spring to Egypt.

Governments, Bad and Fake

There have always been bad governments. For much of the twentieth century, most of the world has had one of two types— the kind that was not good at governing or the kind that was only pretending to govern. When the Soviet Union collapsed, many countries in eastern Europe and central Asia descended into chaos and corruption.

Fewer than two hundred "real" countries have governments and political chiefs who recognize one another in U.N. bodies, international financial organizations, and the like. Even the vast majority of those governments function only in some parts of the countries they represent, and some political leaders don't travel far out of their capital cities for safety reasons. Somalia is but one example of such a country: its government controls a few blocks in Mogadishu, and only with significant help from neighboring nations.

We call these weak, failing, or failed states, and many people recognize that these weak states are a threat to global peace. Most of the governments in these states aren't disruptive, and they rarely have the capacity to threaten their neighbors. They are corrupt and treat their people poorly. Usually, it is criminal organizations that grow up in place of governments—within national borders, or within corrupt bureaucracies—that disrupt global politics. Often criminal organizations operate in parallel with states. Sometimes zones of a state are controlled by local rebels fighting for self-determination.

For example, there are de facto autonomous zones in the Kurdish areas of Turkey and Syria. The Turkish military regularly conducts "raids" within its own borders. The Gorno-Badakhshan region of Tajikistan is another such place, where an ad hoc alliance of Russian military, government officials, and local Pamiri power brokers manage the shipping of a third of all the opium coming out of Afghanistan.

Then there's the northern district of Mali, with Timbuktu, the regional capital, managed by Tuareg rebels and jihadist groups. The Ansar Eddine, a militant Islamist group, managed to cut a deal with a more secular independence movement to chase Mali's army out of the northern part of the country. The rebel groups are factious and hardly law abiding, but they manage an area the size of France and have little to fear from the struggling

Mali government. Boko Haram, a similar group, is trying to isolate a whole region of northern Nigeria from Western contact. When it kidnapped hundreds of girls, no local authority was able to recover the victims or dispense justice. The kidnapping inspired a global campaign, Bring Back Our Girls, that drew as much attention to the region as the Kony 2012 media campaign had.[7] That campaign involved a viral video documentary that reached half of the young adult population in the United States in a matter of days.[8] The activists involved sought to have Joseph Kony, the leader of an armed gang in Uganda, arrested for war crimes and crimes against humanity.

Significant areas of Pakistan are difficult for the country's security services even to enter. North Waziristan, the Punjab, Kashmir, and the federally administered tribal areas are barely governable. In the Philippines, the Autonomous Region of Muslim Mindanao is made up of communities that have voted for a degree of self-rule.

In Peru, the Shining Path was defeated long ago, its leaders jailed or killed and its members pushed far back into the jungle. It still operates within the regions of Apurimac, Ene, and Mantaro, providing security for local drug lords. Some 40 percent of Guatemalan territory is outside the control of the state. Pirates seem to rule the waters off Somalia and throughout the island archipelagos of Indonesia. The Islamic State in Iraq and Syria (ISIS) openly uses device networks to coordinate its communications from hubs in multiple countries. Indeed, these kinds of places are not exactly ungoverned; they are managed by organizations other than governments. The groups that manage these areas sometimes have the power to destabilize their neighbors.

Such spots are the first to fall apart during a political or military crisis. When pushed, central government leaders have to admit they have no control. They focus state resources on the capital cities they inhabit and the areas with national resources that generate wealth. For example, when Syrian rebels started scoring some real victories, Assad gave up his Kurdish region almost immediately, opening a vacuum for the brutal terrorist group ISIS to fill.

The list of limited, failing, or failed states is long. Some governments aren't functioning well, but that suits ruling elites just fine. Governments can be built to fail. Ruling elites extract the resources from the country and know full well that the government won't act against them. The father-son dictators François "Papa Doc" and Jean-Claude "Baby Doc" Duvalier ruled Haiti for a long time as dictators, and did not leave behind a government able to collect and spend taxes, protect property or human rights, or provide for collective welfare.

Not all governments fail entirely, but even having part of a government fail causes problems. When the head of a particular state agency is siphoning off an entire ministry's income to private accounts, the agency won't be able to deliver governance goods. When governments become deeply entwined with criminal organizations, the failure becomes, in a sense, coordinated. It takes only a few corrupt officials to prevent military, police, and justice officials from doing their investigative work. Keeping the state failed becomes an essential service for the criminal enterprise.

Unfortunately, such partially failed governments do provide some public good—often just enough to prevent outright rebellion and keep the criminal enterprise going. That's why drug lords sometimes build roads and keep petty crime in their region under control. They won't allow an environment of competition, so power grabs in other ministries or moves by rival drug lords are always tracked. And there are examples of significant armed forces that provide "governance goods" in the sectors they control: the Tamil Tigers, the IRA in Northern Ireland, and the Kosovo Liberation Army often functioned like the states they were fighting by running local courts and hospitals, and dispensing different kinds of welfare.

The big tragedy is that all these different kinds of places are crowded with people. Almost by definition, slums are among the worst places to live, and about a billion people live in slums.[9] In the years since I was in Haiti, Cité Soleil has grown to at least half a million impoverished souls. Slums are places where water and electricity supplies are meager and economic opportunities are few. Because slums rarely have any political clout, their uncertain status exacerbates residents' challenges. Health services are paltry, and food supplies inadequate. It's not always clear who governs, and even when it is clear, it is rare to find a selfless slumlord.

Mukuru kwa Reuben, for example, began as a labor camp outside of Nairobi, Kenya. It has been around for many decades: now home to well over a hundred thousand people. Politicians have little authority there, and there are no clear property rights for its residents. Rumors of eviction often motivate people, and the outcome can be violent. Not far away, ninety million live in Nigeria's north, a region terrorized by Boko Haram. Almost every country in the Global South has such areas. Some would say that people there live under siege, others that they have the governments they want and deserve. In 2013, the United Nations High Commissioner for Refugees estimated that there were more than fifty-one million refugees living in camps around the world—the most ever.[10] In these liminal states of countries with lousy governments, big data, social media, and the internet of things can have a big impact.

The Dictator's Digital Dilemma

Ruling elites have faced the digital dilemma in many parts of the world, and most are still wrestling with the consequences of their choices. As governments are presented with the option— or need—to regulate the internet of things, they will have to evaluate economic benefits and political risks. I visited Baku, Azerbaijan, in 2004 and met with several activist bloggers. They would speak to me only on the condition of anonymity. Every few months, they would organize a smart mob. At an agreed time, young people would gather in the city's Central Park and open up newspapers. They would sit on park benches around fountains and in circles on the pavement. Then with their newspapers open, they would kick off their shoes all at the same time. The police would come by and get angry with the students. One or two would have to spend a night in jail. The police never knew how to handle such events and never had a coordinated response. Were these events even protests?

When I spent time with these bloggers, they complained that few of their events ever seemed to have much impact on public policy in Azerbaijan. So I asked them why they did it. "These protests are not about toppling the regime," one activist replied after some deliberation. "They are about teaching the regime that the internet now makes collective action possible."

In Minsk, Belarus, protests happen in a similar way. It's not always clear that the protests are, in fact, protests. Belarus's security services are often unsure about how to react. Yakub Kolas Square normally contains skateboarders, promenading couples, and grandmothers herding grandchildren. But on particular days, chosen by some mysterious process of consensus, the square fills to capacity. The park benches are suddenly all occupied. People perch on curbs. Howard Rheingold famously called these gatherings "smart mobs."[11] And at 8 P.M., a chorus of mobile phones goes off: chirps, chimes, and pop music come together to make a cacophony of absurd ringtones.

The police are there, because they too can read the website where instructions for how to pull off this smart mob are posted. Who should be arrested? Can someone be arrested for having her mobile phone go off? More important, is it still a protest if you can't tell who is protesting or why?

Regime response varies, because the digital dilemma is complex. Some people get a few days of jail time for participating in these kinds of protests. Some people get years. Sometimes the police have to make up laws, because it is difficult to prosecute someone simply for having a ringing phone.

Such protests have mobilized large numbers and pose real threats to ruling elites. They attract young people and draw them into protests. Protesters don't always use the methods of past generations: picketing, marching, and chanting. In fact, research shows that digital activism, when it leads to street protests, is usually nonviolent.[12]

Strange protests like the one I attended don't happen just in Azerbaijan. They aren't all as abstract as the ones in Minsk.[13] They happen in Havana. In the final days of Burma's military junta, they happened there, too. What is common is a rising level of innovation in protest strategy. Russia's Pussy Riot does aggressive culture jamming. In Ukraine, the Femen network of young women bare their breasts in public but then talk about pension reform. The Russian art collective Voina painted a two hundred–foot penis on a Saint Petersburg drawbridge to protest heightened security. Ukrainian activists launched a Kickstarter campaign to buy themselves a "people's drone" that would let them watch Russian troop movements in their country.[14]

The internet of things is putting tough regimes into digital dilemmas on a regular basis, because leaders have to choose between two equally distasteful actions. Should they keep the internet on for the sake of the economy? Letting people have mobile phones runs the risk that they will coordinate themselves in some way to talk politics rather than business. Or should device networks be greatly restricted, to preserve political power? So the dilemma persists. Mubarak faced it at the end of the 1990s, but today's aging dictators face it regularly.

For many authoritarian governments, their first digital dilemma was presented twenty years ago. They had to choose between building a communication infrastructure that might connect their national economies to the global information economy, or staying largely offline. As in Egypt's case, being connected offered the potential for economic growth. What dictator would pass up the opportunity to appear modern and make more money for himself and his network of sycophants?

Countries like Malaysia, the Philippines, and Egypt chose this strategy and have faced the political consequences of encouraging widespread technology use. In other countries, such as Cuba and North Korea, ruling elites chose to eschew information technologies. A few countries maintain the social control sufficient to repurpose and reengineer their internet—or to populate their internet with supporters. Only one authoritarian government has the resources to make another choice: China has been building its own internet (more on that later).

The countries that chose to invest heavily in information infrastructure saw some economic benefits and did their best to mitigate the political risks. They hired Silicon Valley firms to develop censorship systems. They passed laws governing who could use the technologies and for what, and they clamped down hard on broadcast journalists. Most countries that chose not to build public-information infrastructure have changed their policies. Cuba has loosened restrictions on mobile-phone use, and the level of political conversation not policed by the state has increased. Myanmar has opened up. Making mobile phones more affordable has been a top priority for civic groups. Only North Korea has managed to keep device networks at bay, simply by not allowing them to be plugged in.

If you were a dictator, would you invest in good information infrastructure? If you don't, you may watch your economy slow down and the world move on. Would you go after people doing creative political protests using digital media? If you don't, you run the risk that symbolic gestures and obscure protest tactics will invigorate your opposition. Authoritarian regimes face the digital dilemma repeatedly: once when deciding whether or not to allow internet access, again when people (invariably) start involving their devices in politics, and yet again when citizens demand access to the latest televisions, phones, and other consumer electronics that constitute the internet of things.

Finding Kibera

The Map Kibera project in Nairobi, Kenya, is one example of how this process has helped a marginalized community figure out its strengths and understand its needs.[15] This act of citizen mapping has made invisible communities visible. And it demonstrates how the internet of things is helping people bring stability to the most anarchic of places.

Primož Kovačič is a soft-spoken Slovenian who left his country in 2007 to work in Africa. He's not sure why he chose to settle in Nairobi, much less in one of its toughest slums. Once there he found a community with immense amounts of economic, cultural, and social capital that had no strong institutions. Kibera is a place where hundreds of thousands of people live. For decades it has been politically invisible because no public map recognized the boundaries of the community, and leaders didn't pay much attention to its needs. Some people call it Africa's largest slum, while the people there call it home.

Estimates vary about how many people actually live there. Some estimate 170,000, others say upward of a million.[16] Whatever the real number is, it is an enormous area and a blank spot on the map until 2009. In fact, many government maps still identify Kibera as a forest. Even Google Maps reveals few details for one of the most crowded and impoverished slums on the planet. By itself Nairobi has some two hundred slums, few of which are on government maps. Some poor districts of the world's megacities, like Nairobi, become what Bob Neuwirth calls "marquee slums": they attract all the big nongovernmental organizations (NGOs) and charity projects.[17] Most of them were not on maps, until Primož Kovačič arrived.

Kovačič decided to help launch a collective project to, at the very least, map the area.[18] Gathering a group of volunteer "trackers" equipped with some basic consumer electronics, including cheap GPS devices and mobile phones, Kovačič and his colleagues "found" Kibera. They identified two hundred schools. They located thirty-five pharmacies. They charted hip bars with late-night dancing, and they found quiet corners where people go to die. They geotagged the sewers, most of which are open, and created layers of data. People from the community were involved in designing the online interface and doing the data collection. They interpreted the data themselves and came to four conclusions.

- They needed more toilets.
- Bad roads meant too many accidents.
- Children had no playgrounds.
- The sewer lines were broken.

Of course, the government saw Kabira as a forest. These new maps demonstrated that it is a community. The community was able to make its own maps for collective needs like clean water and sanitation. Power comes from designing information in a way that lets you act on it.

Focusing on priorities made it easier to act. Within only a few weeks, journalists were covering these problems, small groups of neighbors were working on specific issues, and politicians were alerted. Not only did Primož Kovačič help to find Kibera, but he also helped the community find itself. Social media lowered the cost of collaboration to a point where resource-constrained actors—like Kibera's inhabitants—could agree to solve their problems.

Not everybody was happy about this, of course. Elders and district officers expected bribes. The police didn't like having an organized way of tracking community life that was independent of their authority. Conflicts arose, especially at election time, when the cops claimed that crime was under control. Kovačič told me that the community could even generate their own maps now, showing that crimes were on the rise and demonstrating that police often committed these crimes.

One of the lessons of these mapping projects is that they never just produce maps. Social media make such maps organic, dependent on their contributors for life, and able to expose trends and problems that contributors don't even anticipate. They now have great documentation of the scope of their problems. So Kovačič's next project is to work on a system of finding people and resources to respond. They've clearly mapped the contours of the community and the needs of its people. The next challenge is linking the people in need to the people who can help.

Power comes from designing information in a way that lets others act on it. When I asked Kovačič about the impact of the project, he looked back at the digital map of Kibera. "The important outcome is not the dots on the map," he said. "It's about the social capital and ties that come because of the mapping and after the mapping." In other words, the map was the project. But the volunteer network, civic awareness, and political literacy were the outcomes. Projects like this never really end; they evolve. Software gets repurposed. People build skills and take those skills to new projects. These projects go a long way toward helping people generate information about themselves, especially information that they need the most. In this way, digital media bypasses the state but also bypasses Western NGOs.

Dirty Networks, Collapsing

Slums that can organize themselves are a real threat to dictators. Fortunately, with a few notable exceptions, most of the world's dictators are an aging bunch. This is understandable, because the absence of world wars has meant a prolonged stability that makes both democracies and dictatorships seem more durable. It may seem crass to speculate about state leaders' life expectancies. Given the murderous history of some strongmen who might be on the list, it is not unreasonable to think through the means and implications of their departure, and make a dictators' dead pool.[19]

For example, Saudi Arabia is a constitutional monarchy and an important ally for the West. In 2012, the Saudi Crown Prince

Nayef al Saud died—not unexpected given his age and infirmities. He was the second crown prince to die in a year. The next crown prince, Salman al Saud, turns eighty in 2015, and the Saudi king, Abdullah, turned ninety in 2014. The family tree is complicated, but it is possible to imagine a line of succession. Plenty of authoritarian regimes have no leadership succession plan at all.

Making such a list may seem macabre. But to understand the challenges in international affairs in the next ten years requires thinking about the "predictable surprise" of the demise of these major nodes in the world's dirty networks. Expect crises wherever there is an authoritarian regime with a sick or aging dictator, and ambiguous or nonexistent succession plans. And there are authoritarian regimes where the government has promised elections, and where young, tech-savvy voters will probably use social media to organize against the bald-faced lie of rigged votes. Table 1 identifies the ten toughest dictators who were more than seventy years old in 2015, according to political scientists of the Polity IV Project. Some of these characters have been around for decades, and there are others we could worry about. Among the countries that experts rate as being mildly authoritarian, Zimbabwe's Robert Mugabe is over ninety, Algeria's Abdelaziz Bouteflika is approaching eighty, and Cuba's functioning strongman, Raúl Castro, is over eighty.

Chaos invariably follows a strongman's exit from office—even if he leaves through natural causes. Venezuela's Hugo Chávez was in and out of Cuban hospitals during his third presidential term. In 2012 he successfully manipulated the media and electoral systems to guarantee a fourth win, but died barely three months

Table 1 **Ten Most Authoritarian Leaders over 80 Years Old, 2015**

Country	Leader
Cameroon	Paul Biya
Cuba	Raúl Castro Ruiz
Fiji	Ratu Epeli Nailatikau
Iran	Ali Hosseini Khamenei
Kazakhstan	Nursultan A. Nazarbayev
Kuwait	Sabah al-Ahmad al-Jaber al-Sabah
Laos	Choummali Saignason
Oman	Qaboos bin Said Al-Said
Saudi Arabia	Abdullah bin Abd al-Aziz Al Saud
Uganda	Yoweri Kaguta Museveni

into that term.[20] An election was quickly held, but the country was plunged into a bitter rivalry between Chávez's former vice president Nicolas Maduro and the opposition leader Henrique Radonski. Maduro won, but with just a tiny margin, and widespread protests flare regularly.

This cohort of aging dictators is a major source of instability today. When and how they leave power has an enormous impact on the stability of the countries they rule and the regions in which they have influence. The death of an authoritarian ruler often brings chaos for his subjects and neighboring countries. Even while they are alive, these authoritarian rulers provide important nodes in the global network of criminals.

Even when dictators are young, they tend to generate another kind of problem for the world—they directly support aspiring criminals and other local despots. It's in these failed and limited states that we tend to find pirates, drug lords, holy thugs, rogue generals, and many other kinds of miscreants. They are the key nodes in dirty networks. They represent the real threat to stability. Their networks are also surprisingly fragile when faced with a civic response.

These rulers can operate surprisingly close to home and within parts of the West. In parts of Mexico, drug cartels staff the police force and military with their own people. Indonesia's province of West Papua is rich in oil and natural gas. Immense palm oil plantations have sprung up, pushing aside the rich jungles that have been felled for timber. Foreign journalists have trouble getting in, people have trouble getting their stories out, and the army helps to manage resource extraction for political elites that support the national government. The army basically manages the entire territory.

At any given time, the list of places where anarchy rules contains about a dozen countries. In Somalia, Chad, and Sudan, environmental degradation, ethnic and religious strife, illiteracy, and piracy prevent democratic and civic institutions from gaining much ground. In Zimbabwe, Congo, and Afghanistan, civil war, economic collapse, and kleptocratic governments prevent public organizations from doing much for the common good. Table 2 identifies some of the other places where governance is irregular and governments are incapable. The real challenge in international affairs has become the connection between these disparate places. Leaders in chaotic places sometimes provide hope to their populations. They are led by men who claim authority on the basis of spiritual leadership or military might. They manage small economic empires and they govern, in a fashion.

They govern, sometimes in similar ways to elected politicians in democracies. They collect taxes, dispense justice, and write

Table 2 **Anatomy of Chaos, by Life Expectancy, 2015**

Country	Population (Millions)	Life Expectancy	Symptoms
Afghanistan	29	46	Civil war, drugs, no infra-structure, terrorism
Central African Republic	5	49	Desertification, destitution, disease, terrorism
Congo	68	49	Civil war, massacres, mass rape, looting
Chad	12	50	Desertification, destitution, meddling neighbors
Zimbabwe	13	50	Economic collapse, kleptocracy, oppression
Somalia	9	52	Anarchy, civil war, piracy
Côte d'Ivoire	22	60	Incipient civil war, post-election deadlock
Sudan	43	60	Ethnic and religious strife, illiteracy, tyranny
Guinea	10	60	Destitution, drugs, kleptocracy
Haiti	10	62	Deforestation, destitution, crime
Pakistan	185	68	Coups, drugs, illiteracy, terrorism
Iraq	32	70	Ruined infrastructure, sectarian strife, terrorism

Source: "Where Life Is Cheap and Talk Is Loose," *Economist*, March 17, 2011, accessed September 30, 2014, http://www.economist.com/node/18396240.

the history textbooks.[21] Sometimes they build bridges, maintain small armies, and work with civil-society groups. Sometimes the quality of life for average people living in controlled territories actually goes up—though the distribution of wealth tends to be through sycophants who support the ruling leader. The links between authoritarian regimes involve fuel, loans, and immigration (as well as drugs, smuggling, piracy, weapons sales, human trafficking, and money laundering).

Afghanistan produces at least 80 percent of the world's heroin, and the country's intelligence services have an internal problem with heroin addiction.[22] But in insurgent provinces, the Taliban is actually able to tax the drug lords because it commands key points in the networks of roads leading out of the region. For the drugs to pass, the smugglers need to tithe to the Taliban.[23] Even where the truckers are carrying legal produce, corrupt police can exact some on-the-spot "fines." In Indonesia's province of Aceh, the traffic cops have complex pricing schemes for illegal payments: truckers pay different rates based on the type of cargo and the size of the trucking business.[24]

Sometimes it seems as if smugglers are behaving like governments, and at other times it seems as if governments and smugglers have simply merged operations. Criminal groups and government officials have long collaborated. Even democratic governments have had to collude with criminal gangs to smuggle guns to people fighting authoritarian regimes and to smuggle people out of those countries. Criminal gangs bribe and cajole government officials to turn a blind eye to their operations. Since the fall of the Soviet Union, it has been increasingly difficult for analysts even to tell the difference between criminal gangs and governments in some parts of the world.

These "mafia states" are well-blended organizations in which the highest government officials are also the leaders of criminal enterprises.[25] Those officials dip into the public purse as needed for the defense of the enterprise. Indeed, they work to put official priorities and public policy in service of the enterprise. Bulgaria, Guinea-Bissau, Montenegro, Myanmar, Ukraine, and Venezuela are countries where crime watchers say organized crime and government are inextricably intertwined.[26]

Criminal groups have also started using the latest communication technologies, including software encryption, mobile phones, and the internet, to improve their operations and to find new sources of income. Cybercrime cost the global economy some $113 billion in 2012, according to a leading provider of internet security.[27] This information infrastructure has allowed such a thorough binding between state and crime that the scale and scope of the problem are best thought of as a problem with national security implications and political solutions. Even strong states are increasingly thought to use their mafia connections as an arm of state power. The Russian mafia has been directed to supply arms to the Iranian military and Kurdish rebels in Turkey.[28] The lesson is that, for some countries, foreign policy and criminal aspirations are indistinguishable.

Dirty networks connect extremely poor parts of the world, or connect the poorest communities of the rich world. The number of poor people in fragile states has remained fairly constant for almost twenty years—even considering the changing list of fragile states and the rapid population growth rates of poor communities.[29] Yet their global distribution is changing. The number of poor people in fragile and conflict-affected states has just surpassed the number in stable states. Impoverished countries are not automatically the most fragile ones.

The very device networks that empower dirty networks also expose them to being mapped out. Individuals maintain mobile phones, often several phones in several countries, that allow for geolocation. They are global citizens, too. The *Economist* points out:

> Examples include Somali warlords with deep ties to the diaspora and Western passports; Congolese militia leaders who market the products of tin and coal mines to end-users in China and Malaysia; Tamil rebels who used émigré links to practice credit-card fraud in Britain; or Hezbollah's cigarette smuggling in the United States.[30]

Indeed, of the world's dirty networks, ocean piracy may be the next to collapse. Somali piracy was costing the shipping industry and governments as much as $7 billion a year by 2011.[31] With more than two hundred cases of successful hijackings per year in recent years, a network of naval task forces was established to deal with the problem. The European Union set up a flotilla; NATO provided another; and China, Japan, India, Iran, Russia, and Saudi Arabia coordinated a coalition of warships. These unlikely collaborators have been meeting four times a year to share tactics and intelligence. The Somali government helps, too—it wants to be rid of the pirates as much as any government. These days, Haradheere, the pirate haven, is reportedly devoid of Mercedes SUVs, prostitutes, and kingpins.

Failed states are great at incubating dirty networks. With dictators dying off and the data trail of bad behavior growing, the biggest dirty networks are on the brink of collapse. Under the noses of these aging dictators, and in places that aren't states, you can find a surprising bloom of civil-society groups. These groups are using digital media in creative ways to do the things their governments can't or won't.

The Democracy of Devices

However, we should not be too afraid of the world's dirty networks, and we can be optimistic about their collapse. One obvious reason I've already noted is that the world's dictators—who are important nodes in dirty networks—are an aging bunch. But big data, social media, and the internet of things provide deeper structural reasons to be hopeful. In the next chapter I offer five reasons—phrased as propositions—that I believe are safe propositions for how the internet of things will transform our political lives.

The second is that the internet of things could help people take these dirty networks on, especially if it is configured in smart ways with the wisdom of what we've learned over the past twenty-five years. Fortunately, citizens and lawmakers around the world are already using device networks, in the form of social media and big data, to take down many dirty networks. Every year brings more and more examples of how illicit taxation, drug and people smuggling, and corruption get cleaned up. And the two key forces behind this success are social media and big data.

When people see that their governments are weak, absent, or lousy, they make their own arrangements. Unfortunately, there are many parts of the world where the networks of criminals and corrupt officials are much stronger than the institutions of governance. Increasingly it appears that the best way to battle these dirty networks is with civic networks. The civic networks that equip themselves in smart ways with social media are the ones that perform the best.

People have simply started doing more for themselves, especially through social media and over mobile phones. They have started making their own news. They have started talking about corruption and pollution. They have started coordinating their own health and welfare campaigns. Governments were the primary mechanism for coordinating the public good. People like Eman Abdelrahman, Patrick Meier, and Primož Kovačič demonstrated that device networks could help people build surprisingly agile, effective, and resilient governance mechanisms. Abdelrahman and her friends built their own network of dissatisfied young citizens in Egypt. Meier built his own network of international volunteers for reconstruction in Haiti. Kovačič built his own network of people within Kibera, to serve Kibera.

Wherever and whenever governments are in crisis, in transition, or in absentia, people are using digital media to try to improve their conditions, to build new organizations, and to craft new institutional arrangements. Technology is enabling new kinds of governance. With social media, big data, and the internet of things, people are generating small acts of self-governance in a wide range of domains and in surprising places.

Almost every country in the world now has a digitally enabled election monitoring initiative of some kind. Such initiatives are rarely able to cover an entire country in a systematic way, and they often need the backing of funding and skills from neutral outsiders like the National Democratic Institute.[32] But even the most humble projects to map voting irregularities, film the voting process, or crowd-source polling results help expose and document electoral fraud. Such projects allow citizens to surveil government behavior at a local level, though democracy at the national level isn't necessarily an outcome. Still, the highlight reels of voter fraud can end up online, wearing down bad government.[33]

Ushahidi is a user-friendly, open-source platform for mapping and crowd-sourcing information. These days, there are well over thirty-five thousand Ushahidi maps in thirty languages.[34] In complex humanitarian disasters, most governments and United Nations agencies now know they need to take public crisis mapping seriously. Ushahidi isn't the only platform, though it is one of the most popular because of its crowd-sourced content and community of volunteers. Ushahidi has claimed many important victories in the battle to provide open records about the supply and demand of social services. In doing so, it has taught the United Nations about managing disaster relief with device networks, and has schooled the Russians about coordinating municipalities to deal with forest fires and lost children.[35]

Technology exists in places we don't usually look. Is it providing governance? Device networks haven't entirely replaced government agencies, but people do use information technology to quickly repair broken institutions. Almost by definition, government means infrastructural control. Maps have historically constituted the index to how infrastructure is organized, and are therefore a key artifact of political power. FrontlineSMS, another cell phone enabled self-governance mechanism, helps to improve dental health in Gambia, organize community cleanups in Indonesia, and disseminate recommendations about reproductive health in Nicaragua.

Just because a development project uses device networks in some way doesn't guarantee good governance. When states fail to deliver governance goods, communities increasingly step up, digitally. What we're talking about here is more than service delivery: it is the capacity of communities to set rules, stick to them, and sanction the people who break them. A sovereign state is one that can implement and enforce such policies. When states don't have these capacities, a growing number of communities use digital media to provide services and do so in ways that amount to the implementation and enforcement of new policies. In other words, citizens with device networks are building new governance institutions.

The real innovations in technology-enabled governance goods are in the domains of finance and health. In much of sub-Saharan Africa, banking institutions have failed to provide financial security or the benefits of organized banking to the poor. This stems from a lack of interest in serving the poor as a customer base, but also from a regulatory failure on the part of governments. In some settings, device networks have bolstered social cohesion to such a degree that when regular government structures break down, strong social ties can substitute. If the state is strong but the society weak, information technologies can do a lot to facilitate new forms of governance.[36]

Today, wherever financial institutions have failed whole communities, mobile phones support complex networks of private lending and community-banking initiatives. M-Pesa is a money-transfer system that relies on mobile phones, not on traditional banks or the government.[37] Airtime provides an alternative currency to government-backed paper. Since several countries in Africa lack a banking sector with regulatory oversight, people have taken to using their phones to collect and transfer value. In the first half of 2012, M-Pesa moved some $8.6 billion, far from chump change.[38]

Moreover, people make personal sacrifices to gain access to the technology needed to participate in this new institutional arrangement. iHUB research found that people would forgo meat if it would save enough funds to allow them to make a call or to send a text message that might eventually result in some economic return.[39] A typical day laborer in Kenya might earn a dollar a day, but the value of personal sacrifice for cell phone access amounts to eighty-four cents a week.[40] Two-thirds of Kenyans now send money over the phone. The service is popular precisely because financial institutions are corrupt or uninterested in serving the poor.

Of course, this type of tech-based governance isn't always positive. In India, prostitutes who used mobile phones to organize and protect themselves also talked about the pricing of their services. Over several years, the prostitutes consistently said that their income gets a big boost when they have access to mobile phones.[41] The application of digital media in their business has actually made prostitution a more lucrative career. In many parts of the Philippines, the government is unable to dispense justice in a consistent way, and can't always follow through in punishing those convicted of serious crimes. So vigilante groups equipped with mobile phones and social-networking applications have

organized themselves with their own internal governance structure to dispense justice. Over SMS they deliberate about targets,
determine punishments and delegate tasks. Many countries have
self-organizing vigilante groups like these that deliberate and
unilaterally decide justice when they see the courts fail. Such
groups are responsible for more than a thousand murders in the
Philippines.[42]

These extrajudicial groups have exerted such an important
global force that the United Nations appointed a special rap-
porteur to investigate the problem. The investigator reported
on these networks across a dozen countries. In the Philippines
alone, his reports covered the killing of leftist activists, killings
by the New People's Army, killings related to the conflicts in
western Mindanao, killings related to agrarian reform disputes,
killings of journalists, and revenge killings in Davao.[43] One of the
key findings of such studies is that mobile phones have made it
much easier for vigilantes to meet, deliberate, and act.

Plenty of these technology-enabled governance systems are
stillborn without some kind of state backing. Most of the Congo
is unpoliced, and the government cannot track the movement of
local militias. In the absence of institutions, the Voix de Kivus
network documents sexual assaults, reports on the kidnapping of
child soldiers, and monitors local conflicts.[44] The United Nations
Organization for the Coordination of Humanitarian Affairs, local NGOs, philanthropists, and the U.S. Agency for International
Development study the reports. In this case, the organizers admit that there is little evidence of a governance system taking
root. Reports of conflict are now credibly sourced and appear in
real time, but nobody acts on the knowledge. In order to have

serious impact, most social-media projects need to work in concert with governments.

While some people use social media to provide governance goods, a few use social media to damage or destroy governance goods. Bots can be particularly useful to those who oppose social movements or want to prevent public alert systems from building trust. Bots can also be used to make some public figures appear to be very popular, or to discourage new institutions from growing. Indeed coming under attack is the unfortunate consequence of building successful trust networks that are civic, rather than managed by government or the private sector.[45] Still, citizens and civic groups are beginning to use bots, drones, and embedded sensors for their own proactive projects. Such projects, for example, use device networks to bear witness, publicize events, produce policy-relevant research, and attract new members.[46]

These may seem like isolated examples, but the reason such initiatives are important is that they are contagious. In the past ten years, we've gone from imagining that the internet might one day change the nature of governance to finding a plethora of examples of how this is done. Cell phone companies across Africa, Latin America, and Asia now offer asset-transfer systems, many of which are structured like M-Pesa. International aid can help to prop up a failing state and fund rebuilding operations in a state that has failed.

Of course, people do the hard work of rebuilding. In the new world order, as people see their state falling apart, they pull out their mobile phones and make their own arrangements. Aging dictators may hold together dirty networks, but in many countries there are inspiring blooms of digital activism. Collaborative spirits and problem-solving technologies have been around for a while, but device networks have made creative forms of implementation possible and durable. People are bringing stability to the most chaotic of situations and to the most anarchic places on their own initiative, and with their own devices.

4 FIVE PREMISES FOR THE PAX TECHNICA

Politics, and how people communicate about politics, have changed a lot since the internet became a public resource. Given all the research that has been done on media ownership, political communication, and technology diffusion, what safe generalizations would be reasonable premises about the role of the internet in political life so far?

1. Political leaders, governments, firms, and civic groups are aggressively using the internet to attack one another and defend their interests.

2. Citizens are using the internet to improve governance, and the success or failure of a government increasingly depends on a good digital strategy.

3. People are using the internet to marginalize extremist ideas, and authoritarian governments lose credibility when they try to repress new information technologies.

4. People are using digital media to solve collective action problems.

5. People are using big data to help provide connective security.

The Romans and British built stable infrastructures and prosperous societies by linking widespread territories through networks of roads, family ties, and trade routes. The technologies of the internet of things will have a similar role, and are already providing some palpable conduits for political power. Our internet has features that many privacy advocates dislike, but the problem-solving capacity and human security benefits of responsibly handled device networks trump the risks. The tough project—in the years ahead—is getting security agencies to behave responsibly.

In this chapter, some historical perspective about technology diffusion over the past twenty-five years helps turn the observations of the previous chapters into concrete premises. These five safe premises about how we have used the internet for political life can help us—in the subsequent chapter—envision the consequences of the internet of things.

Learning from the Internet Interregnum

Alexis de Tocqueville was one of the first people to investigate political life in a systematic way. While most Americans know him for his observations on the early stages of their democracy, he also sought to understand the nature of revolution. Living through the tumultuous upheavals of 1848, he wrote in his *Recollections* that big political changes were the outcome of a myriad of causes.[1] Louis Philippe, France's last king, was chased out of the country in 1848 for what Tocqueville described as "senile imbecility." The most important causes of political change, he concluded, were clumsy leaders and unfortunate mistakes.

Still, Tocqueville argued that chance events and idiotic leaders are only "accidents that render the disease fatal," and that while political change is nearly impossible to predict, it clearly happens only at particular moments. Understanding the particularities of the revolutions of 1848 or the collapse of East Germany involves recognizing timely trends and comparative contexts.

The world is full of complex problems. Governments fall apart, terrorist cells persist, and, in any given year, a handful of countries suffer from genocide, internal warfare, and human-rights violations. We need to worry about nuclear proliferation, high population growth, and migration pressures. Entire regions are disrupted by debt crises, viral diseases, and breakdowns in our energy supply. Humanitarian crises brought on by environmental degradation, persistent poverty, and debilitating malnutrition affect millions of people each day. Given the complexities of all of these problems, why should we worry about the internet of things? How does understanding technology diffusion help to solve complex problems and explain political change?

No singular cause determines social outcomes: there is always an interplay of causal factors. This makes it tough to learn from the causes and consequences of technology diffusion throughout history. Important events and recognizable causal connections can't be replicated or falsified. We can't repeat the Arab Spring in some kind of experiment. We can't test its negation—an Arab Spring that never happened, or an Arab Spring minus one key factor that resulted in a different outcome. We don't have enough large datasets about Arab Spring–like events to run statistical models. That doesn't mean we shouldn't try to learn from the real events that happened. In fact, for many in the social sciences, tracing how real events unfolded is the best way to understand political change. The richest explanations of the fall of the

Berlin Wall, for example, as sociologist Steve Pfaff crafts them, come from such process tracing.[2]

We do, however, know enough to make some educated guesses about what will happen next. And, indeed, our experiences during the internet interregnum reveal five premises for how the internet of things will have an impact on global politics. I opened the chapter with basic statements of how device networks have had an impact on political life. Now let's expand on each of those five premises.

First, it is safe to say that powerful political actors will use the internet of things as a weapon. They already use the internet to spy on one another, and they use it to manipulate public opinion. They defend their interests. Sometimes, they aggressively attack one another. Many of the key technologies that make up the hardware and software that we use today were developed by the U.S. military during the Cold War. Only recently have political actors figured out how to take full advantage of these technologies. Fortunately, there are good reasons to expect that cyberdeterrence will bring about the kind of peaceful stability we've seen in the past when political actors have maintained the balance of power, refrained from destructive attacks, and used their new weapons mostly for deterrence. The result of such stable positioning is a dynamic balance of power.

Second, when the modern state fails, the internet of things will provide governance. Or more accurately, people who live in places where governments collapse or fail to provide particular services use information technology to coordinate their own governance. These days, we also get immense amounts of data from failed states. So while failing or failed governments previously caused all sorts of problems for citizens and neighboring countries, now communities increasingly respond innovatively on their own, and more outsiders respond with more effective help. The result is often stability rather than anarchy.

Third, new information technologies undermine radical ideologies, and the internet of things will contribute to this trend too. Radical opinions simply don't last very long on the internet. Or they persist only in the dark corners of the internet, before they get pushed off into the margins by people who do fact checking and present reasonable alternatives. When political elites manage to sequester their supporters via information choke points, ideological biases can last longer than they should. For the most part, people use information technologies to take the hot air out of ideologies. There are still a lot of bad people with bad ideas. On the whole, a world without radical ideological differences is a more stable world. Only one important ideological divide is worth talking about. Should the internet of things be open or closed?

Fourth, social media help people solve collective-action problems, and the internet of things will greatly deepen our ability to coordinate collective action. The number of stories in which information technology has helped community leaders address local problems grows every year. The kinds of problems that we can solve are diverse, from water scarcities and pollution to public-health needs and human rights. Some of these are tackled in niche hackathons, some of which are very issue specific.[3] Others get addressed through social media, which puts transportable solutions into new community contexts. The result is that even some of the most intractable collective-action problems, sometimes called "wicked problems" due to their sheer maddening messiness, are being solved through new networks of information and creative-coding skills.[4] By helping us overcome these challenges, the internet of things will help engender social stability.

Fifth, big data is providing us with collective security. The internet of things will make the data resources about our attitudes and behaviors so enormous as to be beyond the ability of the social sciences to interpret and make use of much of the material. Immense amounts of data about criminal activity, mafia strategy, drug lord schemes, and terrorist plans have already made it easier to protect people. Yes, some of the ways national security actors have obtained this big data are grotesque violations of the public trust. Reworking the ways in which we can oversee the collection of data is a key priority. Respectfully collected, big data helps us provide for our own security. Let's examine each of these premises one by one.

First Premise: The Internet of Things Is Being Weaponized

The funders and founders of early internet architecture were U.S.-based military and research organizations looking for ways to distribute the command and control functions of military assets over wide areas. The internet has always been a powerful tool for surveillance and social control, even when private entrepreneurs started generating innovative new hardware and applications to attach to the internet. Device networks have been used by combatants in many different kinds of political competition and military conflict: border skirmishes, preemptive attacks, assassinations, old wars, new wars, propaganda wars, espionage, and coups. Whatever devices come embedded with sensors, power packs, and network connections, someone will try to weaponize them and attack opponents.

Authoritarian regimes have invested significant resources in attempts to prevent their populations from being exposed to information from more democratic countries. The raft of NSA surveillance scandals reveals that even democratic countries use the internet as part of their national security arsenal. The internet has been "commandeered," in the words of one prominent privacy advocate, Bruce Schneier.[5]

The internet has become not just a weapon in the world's great political battles. It has become the weapon for ideological influence, and careful use can mean the difference between winning and losing. Device networks have proven useful in the short, medium, and long game of politics. Digital media have become the most important offensive weapons, as they are where political battles play out, and their successful use goes a long way toward ensuring victory. In some ways, military strategy has been subsumed by media strategy.

Even before Edward Snowden exposed U.S. surveillance activities on the global internet, many people were concerned about the malware on device networks built by China's largest telecommunications firms, Huawei and ZTE. Australian, Canadian, and U.S. officials have all objected to technologies from these firms on the grounds that the hardware would allow Chinese spies backdoor access into whatever networks they get plugged into.

We don't know much about the malware Chinese firms are putting in the devices they sell. There appears to be enough classified evidence for multiple governments in the West to justify import bans, block development contracts, and talk openly about the national security risks of hooking Chinese equipment up to Western networks. This doesn't mean Huawei and ZTE are short of customers. Many countries in the developing world, such as Zimbabwe, sign long-term infrastructure deals with China.

Networked devices and the data trails they generate can help police track criminals and allow militaries to target individuals. Dzhokhar Dudayev, the Chechen separatist leader, was traced by Russian security services when he used his satellite phone. Mobile phones with embedded bombs have carried out assassinations for the Israel Security Agency.[6] It was the suspicious absence of networked devices and information infrastructure linking up an expensive mansion in Abbottabad, Pakistan, that betrayed Osama bin Laden's last hideout.[7]

Unfortunately, this also means that governments increasingly use networked devices to track the people protesting against them. During Ukraine's Euromaidan protests in the winter of 2014, the government there exercised its technology clout and pulled off what is probably the first example of geotagged propagandizing.[8] After several protesters were shot dead by police, the government identified all the mobile phones connected to cellphone towers in the conflict area and blasted out the message "Thank you special forces of Ukraine for saving the capital" and signed the message "Citizens of Kiev."[9]

This enraged even more people, because it was easy to verify that the protests had been mostly peaceful up to that point. It was clear that the state was getting desperate, and reaching into a very personal technology and using it to manipulate public opinion. When protesters responded with more energy, the next geotagged message from the government was more ominous: "Dear subscriber, you are registered as a participant in a mass disturbance." This was probably the first example of a troubled government using networked devices to geolocate people in real time and go after them with a counterinsurgency campaign.

Most of the recent popular uprisings have exposed the ways that repressive regimes use networked devices as a weapon against dissent. Civic leaders from around the world condemned the company Nokia Siemens Networks for selling its Lawful Interception Gateway device to Iran.[10] The equipment allowed the Iranian government to put down the 2009 Green Revolution by blocking internet and cell phone traffic. Ericsson's equipment could have been used in similar ways against civic leaders in Belarus during protests there in 2010. A Boeing subsidiary sold powerful net inspection technology to Egypt's state telecom in 2011, equipment used to examine the communications of the country's bloggers and civic leaders.[11]

But these days, devices on the internet of things can be targets as much as people can be targets. The Stuxnet virus is the earliest, most dramatic example of how a sophisticated military application can achieve a security objective by attacking devices. It was designed to make Iran's nuclear enriching uranium centrifuges spin out of control.[12] The technical specifications are meaningful only to the handful of engineers who built that equipment, the specialists who wanted to use it, and the designers of the virus. It installed malware into memory block DB890 of the Profibus messaging bus of Siemens S7-300 centrifuges that used a variable frequency motor built by either Finland's Vacon or Iran's Fararo Paya, if those systems cycled between 807 and 1,210 per second. The malware periodically changed the rotational speed, from as low as 2 cycles to as high as 1,410 cycles per second. But it also installed a rootkit that misled monitoring systems. In other words, the virus made a specific kind of equipment built by a particular company—if it used one of two distinct motors operating at specific speeds—stress itself enough to break down. It hid the whole process from Iran's nuclear engineers and dealt a serious blow to the country's nuclear program.

Digital networks are also the staging ground for domestic political battles. In Turkey, the military attempted what the country's journalists called a "coup-by-website."[13] Its government continues to wrangle with Twitter.[14] In the first instance, elites in Turkey were unhappy about the likely election of a mildly Islamist president, Abdullah Gül, in 2007. Publishing an online memo reviewing the military's obligations for protecting Turkish secularism and its options for Gül's election created a national crisis. Gül eventually won the election and came out ahead—with many of his military challengers going to jail.

When Erdoğan banned Twitter and YouTube before local elections in 2014, there was a lot of online outrage. His party was easily returned to power. Shutting off vocal opposition at the right time still works, but it's getting harder and harder to do.

These days, media targets in the West—and particularly the United States—are incredibly valuable. Taking down the New York Times' website is one of the tactics of modern warfare.[15] In August 2013, a hacker group loyal to Syrian president Bashar al-Assad known as the Syrian Electronic Army took down the Times' site and also attacked Twitter and the U.K. site of the Huffington

Post. It was not the first time the group had attacked a Western media organization, but it was the first time that it was successful in denying online service for the *Times.* Indeed, the *Times* has become one of the most valuable targets for anti-U.S. hackers.

In emerging democracies, the police sometimes weigh in to help a government survive an election. In the 2006 elections in the Democratic Republic of Congo, the security services tried to shut down phone numbers used by opposition leaders. In 2009, Colombian security services used U.S. wiretapping technology from the war on drugs to surveil the government's political opposition.[16] Modern militaries have long been purposeful about developing a media strategy. Napoleon famously quipped that "four hostile newspapers are more to be feared than a thousand bayonets." Militaries are now quick to launch social-media campaigns, and increasingly they begin their media offensive minutes ahead of action on the ground.[17]

Almost every variety of war, conflict, and competition now has its cyberequivalent. The substantive threat—that the internet of things will be used for attack—is only going to increase, in part because of the potential to make such attacks effective but tough to source. Espionage is always a threat, but cyberespionage seems to be the new, more productive strategy for Chinese businesses, the Russian mafia, and the Syrian government.

Digital networks are still relatively decentralized tools: so many different kinds of actors can use the new weaponry. Certainly hackers and technology ideologues such as Anonymous use the internet to attack political leaders and organizations they believe aren't behaving well. And groups like the Tactical Technology Collective teach civil-society groups and democracy advocates how to fortify themselves with information technology.[18]

Governments and businesses use the internet to spy on one another, and the United States—and its allies—use the internet to spy on everyone.

Still, it's getting ever harder to control the civilian application of technology. For several hundred years, the military either invented, cultivated and developed, or first applied technology. The military is still an important source of innovation, but the world is now rife with examples of technologies that are being used by civil-society groups in creative ways while governments think about regulations. Sometimes these governments give up on regulation and try for bans.

In 2013, Human Rights Watch used satellite images to show the abusive Nigerian Army wreaking havoc on the town of Baga.[19] When the military raided the town, searching for Islamist supporters, it left behind a swath of destruction. Local community leaders claimed that more than 2,000 homes had been burned and 183 bodies identified after the military raid. Human Rights Watch corroborated the account with satellite images, identifying 2,275 destroyed buildings and another 125 buildings severely damaged. In 2014, the organization combined satellite imagery, public photos, and photos released by ISIS militants to reconstruct the execution of between 160 and 190 men in a field near a former palace of Saddam Hussein.[20]

Protesters now use drones.[21] Peaceniks mine Twitter for crisis data. The lesson isn't so much that information technology can be used by political actors as that the innovators and developers of information technologies have less and less control over who uses these innovations, and for what ends. Only a few armies and navies could ever equip their forces with cannons. Now, both militaries and peaceniks can equip their members with digital media. Civilians have access to these tools, and they don't always act in concert with their militaries.[22] Many civic groups have creative digital media projects that inform their members and make policy makers think in new ways about old problems.

The aforementioned Tactical Technology Collective helps civic groups and social movements develop sophisticated and secure communications strategies.[23] FrontlineSMS, discussed in Chapter 3, is an open-source text messaging service used by nonprofits for distributing information about politics, health, and welfare.[24] The Mobilization Lab is used by environmental groups to experiment with new ways of reaching their supporters and coordinating their campaigns.[25]

The internet of things—and the ability to manipulate devices— is the defining feature of modern political conflict. Countries spend ever more money on information infrastructures: on ways to surveil their people and disable enemy infrastructure. The internet of things could not have been built without the entrepreneurship and inventiveness of technology firms. Nonetheless, our surveillance state also could not have been built without that inventive industry.

Second Premise: People Use Devices to Govern

The state appeared as the dominant political form some five hundred years ago. As a way of organizing resources, states were good at building infrastructure. But for the first time the major infrastructure for social cohesion is not owned, managed, or even closely regulated by the state. Now, the information infrastructure that connects us is not monopolized by a particular political actor, or even by a particular kind of political actor. It exists and grows independent of any single political, economic, and cultural actor. When it comes to political power, people increasingly use technology to supplement or supplant government.

When governments do succeed at something these days, it is often because they have used information technologies to serve citizens in creative new ways. Governments have lost the exclusive power to frame current events. When governments fail, people repair their institutions with digital media. Monterrey's public alert systems and Kibera's mapping project, discussed in Chapters 1 and 3, are examples of how this can work.

James Scott is an anthropologist famous for demonstrating how much political power governments got from simply being able to define a public problem.[26] For example, a government derived its power from being able to draw maps because that made everyone think about national borders. Today, anyone with internet access can create, redraw, and share a map. And people have been using Ushahidi to draw entirely new maps that serve community needs over those of political elites. Internet users run their own public opinion polls when the government won't call an election or the election is rigged. People use mobile phones to report a crisis, humanitarian or political.

Every modern political crisis comes with clumsy attempts to control the way events and facts are disseminated online. When governments are on the verge of collapse, their leaders quickly figure out that they need to manipulate digital media to save themselves. Some countries, like Algeria, find that in a crisis they just don't have the technical skills to mount much of a digital counterinsurgency strategy. When the Arab Spring arrived on Algeria's doorstep, it simply had no centralized management system to coerce the country's internet and mobile-phone companies to support the regime. Other countries find that they have to rely on outside firms to do their bidding.

In Egypt, Mubarak required the assistance of London-based Vodafone to shut down national networks. Yet cutting off twenty million internet users and fifty-five million mobile-phone users only ratcheted up the political tension. The OECD estimates that this move cost the Egyptian economy $90 million a day for five days. Cutting off connections between friends and family— the majority of whom were not in the streets protesting—compounded public anger.

If a government is working at peak performance, it can accurately frame a problem and help all the stakeholders prioritize solutions. If it isn't working well, digital media allow stakeholders to investigate and propose alternatives. The internet of things will make it harder for a regime to control "things" attached to networks and choke off information flows. Moreover, stakeholders will have ever more ability to do their own research and craft their own policy proposals.

People use the internet to compound attention on poorly performing governments. A group of researchers recently did a broad evaluation of this process. Merlyna Lim, for instance, found that authoritarian Egypt failed to respond to the communities of opposition that coalesced online well in advance of 2011, while Zeynep Tufekci and Chris Wilson illustrated that social media removed the *disincentives* for people to join Tahrir

Square protests in early 2011.[27] Catie Bailard showed that internet use predicted cynicism during a Tanzanian election.[28] Jonathan Hassid demonstrated that Chinese bloggers lead in the framing of issues when the ruling political and media elites do not appear to be acting responsibly; and Sebastián Valenzuela, Arturo Arriagada, and Andrés Scherman's study of Facebook use in Chile illustrated that social media can effectively mobilize those who are not already involved in political activism.[29]

Sometimes governments do figure out ways of using technology to improve themselves. Tech-savvy governments can often expose and stop corruption. In the Nigerian state of Bayelsa, a new biometric verification system of fingerprinting public employees and matching them with employment records confirmed twenty-five thousand legitimate employees but four thousand illegitimate ones. Most of the fraudulent employees were in the finance department. The local office of the national electoral commission had extra employees, including seventy people who claimed to work there but didn't. An elected school board member employed ten members of his family—including underage children—in the board. In 2009, new systems like this allowed Nigeria to reduce its state salary budget by 20 percent. Similar smart databases cut the state procurement budget by 24 percent. Automated administrative systems did more to fight corruption than awareness campaigns and legal threats.[30]

This doesn't mean that governments can't be innovative online. When states succeed at serving their publics these days, it's increasingly because government bureaucrats have figured out creative ways of putting technology to work for the public good.

Research shows that most civic projects that use information technology in creative ways need to be designed in concert with government. Markets can also serve as government mechanisms, and there is much evidence that device networks can help get rid of discrimination in markets. In India, clear price signals over mobile phones and dedicated apps have brought down prices and raised profits for fish markets in Kerala and soya beans in Madhya Pradesh.[31] Civic projects that are totally independent of government legitimacy often fail; on the flip side, government projects that have little or no buy-in from civil-society actors often fail.

E-government services can bring transparency to procurement processes and make services more accessible to citizens.[32] Governments do all sorts of information-intensive tasks, well beyond service delivery. A government that doesn't use information technology well loses its legitimacy quickly. Keeping track of ballots on election day, noting which ships are in the port, who's in jail, and who needs a driver's license are all logistical challenges. Effective technology use has come to define good governance, whether creative initiatives come from people or their governments.

Third Premise: Digital Networks Weaken Ideologies

On July 9, 2008, Iran wanted to show the world its new mobile missile launchers. Leaders shared their triumph through high-resolution photos of the test site. The world's major media outlets carried an image of four missiles blasting into the sky. The image was reproduced the next day, first on the Agence France-Presse webpage, then on the front pages of the Los Angeles Times, the Financial Times, the Chicago Tribune, and several other newspapers, as well as on BBC News, MSNBC, Yahoo News, nytimes .com, and many other major news websites. Somehow, a different image was sent to the Associated Press—an image with only three missiles successfully launching—one missile had actually broken down and failed to launch.

Many of the world's media outlets published retractions. Journalists wrote apologetic essays about how technology had made it too easy for manipulative regimes—such as Iran's theocracy—to doctor images. The important lesson here is not that propaganda experts in the regime used Photoshop to make their country look more powerful. The lesson is that the manipulation was caught, by 3 P.M. on the East Coast of the United States, the very next day.[33] Images are powerful because they can bolster or dissolve political authority.

Digital media have not only been useful in killing off ideological propaganda, they have allowed democracy advocates to keep political memes alive and to make their issue go viral. Where ideologues and their ideologies do find traction and audience, it is usually because the messengers have been especially effective at using technology to promote their message and to keep their followers corralled, not because the rhetoric or ideas are compelling or sensible. This means that rival messengers, with better technologies, can get the upper hand in a political battle.

In China, one of the most provocative political images around is still that of a man standing in the way of a tank in Tiananmen

Square. The image, taken June 5, 1989, has become one of the most iconic images of political resistance. It has been cropped and retouched many times, but the Chinese have well-developed image analysis software that detects and removes the image whenever it pops up in the country's digital traffic. Images of Chinese protest events are tough to find on the national search engine, Baidu. Yet careful editing has kept the image alive, most recently by a mash-up that replaces the tanks with large rubber duckies.[34] Other versions involve a Lego figure standing up to Lego tanks.[35] Such visualizations keep the hope and spirit of civil disobedience alive and out of the automated surveillance net.

These are only examples of how digital images can undermine political ideology. The impact of new digital networks on ideology is bigger than just these stories. Indeed, the ideology of technology is trumping all others. There have been no truly new ideologies since the end of the last world order, and the closest thing to a new ideology is the ideology of technology itself.

As a concept, an ideology can be defined many ways. Among the best is the understanding that ideology is "meaning in the service of power."[36] The dream of a truly wired society, with tech-savvy citizens and responsive e-governments, is part of such an ideological package. But this dream about what an information society should be, widely promoted by government and industry, also serves the interests of the businesses and politicians who deliver on this version of modernity. The internet of things is also becoming a kind of ideological package: internet use and networked devices have become deeply associated with our notions of modernization and economic growth. Popular imagery about the use, speed, and sophistication are pervasive, with some technologies becoming iconic in myth and symbol, and inspiring almost religious fervor.

Manuel Castells has called this "informationalism."[37] While I have argued that there have been no new ideologies since the collapse of the Soviet Union in 1991, the ideology of internet-led growth might be the exception. Our infatuation with the internet drove a dot-com boom in the economy, it inspired a re-thinking of global development priorities, and it remains a per-vasive Western export: the notion that information technologies can fix most problems. The economic buzz around technology startups has given entrepreneurs clout in culture and politics. In some countries in the Middle East, for example, this has meant that governments are trumpeting entrepreneurship and innova-tion over traditional Islamic values.[38] Today, information tech-nology is the most important tool for servicing power.

The internet has a strong record of marginalizing partisan-ship, radicalism, fundamentalism, and extremism in social net-works. Increasingly, the outcomes of both domestic and interna-tional political battles seem shaped if not determined by patterns of digital media use. Traditional ideologies have lost the power to frame events, and radicals in many countries have either had to soften their message, learn to manipulate the internet, or be socially marginalized. And increasingly, it is through decisions on technology policy that governments reveal the true extent of their commitment to democracy.

Ideologues spend a lot of time thinking about image. While powerful images can support an ideological perspective, the wrong images can deflate ideological claims. So technological resources determine domestic political battles. Political candidates in emerging democracies use digital media to raise funds, rally supporters, and outmaneuver opponents in policy debates. A growing number of upstart leaders and new political parties manage to achieve their political goals by manipulating the internet of things.

In authoritarian regimes, where elections are a farce, such rigged events have become especially sensitive moments. Many of the most violent confrontations between dictators and their opponents have come because civic leaders used digital media to document the depth and scale of electoral corruption. Even in China, where the Communist Party has no tolerance for open dissent at its executive levels, that same Party has been allowing—some would say encouraging—two kinds of political activism in local politics. Communities that use the micro blogging with Sina Weibo or instant messaging service Tencent QQ to rail against local corruption or environmental concerns seem to get the Party's attention. People can use social media to vent, a little, and at certain authorized targets. Doing so reaffirms that the Communist Party of China is ultimately in charge.[39]

In democracies, smart use of technology increasingly gives a political party the upper hand at election time. At this point, good examples of this go back several years. Student rallies organized rapidly by SMS toppled Philippine president Joseph Estrada in 2001, when protesters gathered en masse. They were summoned together by a single line passed from phone to phone: "Go 2 EDSA [an acronym for a Manila street]. Wear Blck."[40]

President Roh Moo-hyun ushered in a new era of politics in South Korea, but he would not have been elected without the help of the internet and SMS. Back in December 2002, conservative mainstream media favored his rival Lee Hoi-chang to win the election, especially when a former rival who had endorsed Roh unexpectedly withdrew his support on the eve of election day. Roh's young supporters launched a massive last-minute campaign, sending off emails and text messages to 800,000 Roh supporters to remind them to vote.[41]

In Spain, the Madrid bombings had direct political consequences as a result of communication newly enabled by technology. The ruling conservative Popular Party had been aggressively defending its close ties to George W. Bush's war effort. When the terrorist bombs went off in March 2004, a wave of popular dissent cascaded by SMS through the electorate, faster than government spin doctors could handle. The overwhelming viral campaign cost the government its position of power. When public outrage goes viral, leaders in democracies are especially susceptible.

In more and more elections, political victory goes to the most tech-savvy campaigner. Ideological packaging seems secondary. To be a president or a prime minister you still need an impressive party machine, a good smile, and at least a few decent policy ideas. These days, an impressive party machine is one that uses social media to create a bounded news ecology for supporters. It mines data on shared affinity networks, and otherwise mobilizes voters on election day.

Research on elections in Brazil and Malaysia demonstrates that one of the most important statistically significant predictors of actually winning a parliamentary seat—especially in lower houses—is being a tech-savvy candidate.[42] Having a Twitter feed and an interactive website helps connect with voters. And online search habits leading up to an election help predict which candidates will win.[43] Around the world, being a modern politician means more than having a decent website. It means being able to work with the information infrastructure that young citizens are using to form their political identities.

Ideologies, like governments, have lost much of their ability to exclusively and comprehensively frame events. Indeed, the claim of Francis Fukuyama's "End of History" argument is that there will be no more great ideologies because capitalism has triumphed over all of its rivals. While it may be true that there have been no great ideologies since the arrival of the civilian internet, it's also true that when there are ideological battles, they happen online. What makes an ideology successful is its ability to prevent followers from being aware of the way public issues are being framed. With a worldwide network of watchers, trying to doctor photos or censor unflattering images is quickly met with a corrective from somebody in the network.

High-ranking Chinese officials certainly feel this way. Liu Yazhou, political commissar of the University of National Defense, published an article in the *People's Liberation Army Daily* arguing that today's internet has become the main battlefield for ideological struggle. "Entering the new century," he wrote recently, "whoever controls the internet, especially micro-blog resources, will have the right to control opinions."[44] The Party is aware that political conversations over social media have real-world consequences and can provide a metric of public opinion. Senior officials get exclusive access to social media sentiment analysis through the Party's media research team. One Chinese pollster blames a 10 percent drop in confidence in the Party to the rapid spread of microblogs.[45]

When moderates and ideologues are given equal access to digital media, people tend to use social media to marginalize extremism, hate speech, and radical ideas. In part, this is because digital networks are ultimately social networks. On a personal level, we often don't like experiencing "socialization" because it can mean embarrassing correctives to our bad behavior. The pressure to conform is rarely a pleasant thing to experience. Socialization also means that dangerously violent behavior, and the ideas that might foment such behavior, get stigmatized. The problem, of course: not everyone has the same degree of internet access.

The research is growing on how social media marginalizes bad ideas, and it is based on varied levels of analysis. Sociologists have found that digital media have several positive long-term consequences for users. Over time, people develop increasingly sophisticated search skills.[46] They tend to become more omnivorous with their news diets.[47] And there is evidence that they become more tolerant of ideas and opinions that diverge from their own.[48] Research suggests that digital networks moderate political opinion, and this is because the average person is moderate.

The internet has grown up along social networks. For better or worse, socialization works. Over time, people with extreme, radical, and disturbing ideas either moderate their opinions or find themselves marginalized in their networks of family and friends. Unfortunately, the corollary of socialization through digital networks is that extremists may have an easier time finding others like them. So pushing violent extremism out of mainstream political conversations may make some small networks of extremists seem to grow and become more resilient. But as we'll see, this process also makes it easier to track and disable those networks when they become a threat.

The process of socialization over digital networks doesn't have a positive impact just on individuals; it can be observed in political discourse as well. For example, experiments with online news rating systems show that social influence accumulates positively. We are social animals who tend to herd positively and create ratings bubbles of approval. The corollary is that negative influences get neutralized by the crowd. We tend to put a little effort into making sure a negative news story is deserved.[49] This is why political parties usually compete for the middle. And not just during election time, either. Even autocrats have to maintain a balance between hegemony and public appeal.

The Arab Spring may be the best recent example of how moderate, digitally activated citizens coalesced into viable opposition movements. Only after several weeks of protest in Tunis and Cairo did Tunisia's and Egypt's Islamists decide to join protests against the authoritarian, secular governments. Islamists eventually formed governments in both countries during open and fair elections. But both groups of Islamists had to moderate their messages—they had to give up advocating for polygamy and the traditional punishments mandated by the Koran. When the government formed by the Muslim Brotherhood in Egypt drifted too far, and excluded too many secular groups, civil disobedience erupted again and the government was tossed from power.

In Egypt, networks of democracy advocates proved their resilience by activating twice: they toppled a secular authoritarian regime that had lasted for thirty years, and then they toppled the government formed by the Muslim Brotherhood after a year. Digital media helped disaffected youth defeat radicalism and push Mubarak out. Then it helped them push the Islamists out.

The Muslim Brotherhood in Tunisia and Egypt are no longer the only viable parties in their reborn countries. Both must continually address concerns about their radical roots. Terrorist groups might find some safe harbor online by claiming tiny corners of the internet for recruiting impressionable members and coordinating activities. Islamists who hope to participate in political life, on the whole, are having a hard time. In Indonesia, Malaysia, and Turkey, Islamist political parties have had to moderate their message, and they often get punished by voters when they try to introduce radical legislation.

Indeed, there is a growing interest in using the network effects of digital media to consolidate radical groups more aggressively: dedicated social-network applications to support ex-neo-Nazis and ex-terrorists,[50] mainstream civic groups and think tanks made up of former members of extreme groups.[51]

When someone in a migrant community comes from overseas and acts badly in the name of his faith or some perceived injustice in his homeland, news headlines tend to blame the internet for allowing radical ideas to spread among immigrants. But research suggests that while immigrant communities do use digital networks to construct new communities and stay in touch with their communities of origin, they tend to use it to defeat nostalgia about the conditions they left behind.[52]

Even more than declared ideology, a country's technology policies reveal its true political values. While political scientists wrestle over the meaning of democracy and authoritarianism—and which countries are the best examples of each—technology policy has come to be the most revealing aspect of a regime's priorities. When a government spends money on technology initiatives that make the business of governance more transparent, we celebrate. When a government decides that it needs to be doing more censorship and surveillance, we rightfully worry. In recent years, we find authoritarian regimes doing the former and democracies doing the latter.

Democracies don't always get information policy right either. A powerful mobile-phone surveillance tool, the Stingray device allows a user to spoof a mobile-phone tower. Such technologies can be used to track terrorists in India, or drug dealers in Los Angeles.[53] Warrantless wiretapping is a more high-profile concern in the United States, so the country's civil liberties groups quickly linked the use of the Stingray to privacy issues.

Singapore controls its journalists at election time so as to ensure the governing party returns to power, but puts its taxation and spending records online. Canada aggressively surveils its citizens—even travelers using airport wifi.[54] Which country is more open?

Videos from abusive dictatorships consistently expose the attitudes of ruling elites, and we might expect these countries to have active surveillance and censorship programs. Even liberal democracies have been running aggressive programs with relaxed public oversight. Google transparency reports show that requests for information are on the rise, and most of the requests come from within democracies.[55]

For modern dictatorships, all the new devices being connected to the internet present a real challenge. Some authoritarian regimes may run honest elections administratively but invest in social-media strategies that guarantee electoral victories. Russia makes significant investments in video equipment for its polling stations during referenda and elections. Their leaders decided that video evidence of fraud is not admissible in fraud complaints.

Today, democracy is a form of open society in which people in authority use the internet for public goods and human security in ways that have been widely reviewed and publicly approved. Democracy occurs when the rules and norms of mass surveillance have been developed openly, and state practices are acknowledged by the government.

Information policy has not only come to define what kind of government a country has; the political decision to disconnect information infrastructure now delineates a regime on the edge of collapse. Net watchers report instantly when packet switching through a nation's digital switches stops and the country "goes dark." Public protests in an authoritarian regime can be a sign of political instability. A defining feature of political, military, and security crisis is the moment when a ruler orders the mobile-phone company and internet-service providers to shut down. Going dark has become the modern mark of a regime in crisis, and the indicator that a state is close to collapse. Contemporary authoritarianism, democracy, and state failure are now defined by technology use.

Moreover, the way a political group treats digital infrastructure has also come to define serious insurgency. Nigeria's Boko Haram figured out that digital media were being used to track its members, so it began taking down cell phone towers.[56] Afghanistan's Taliban takes down cell towers for fear that they help unmanned probes track their leaders. Lebanon's Hamas has its own hard lines, which it defends in times of chaos. Even the Zapatistas knew that the first step in their insurgency was to disconnect the information infrastructure leading out of Chiapas. Now, Hezbollah owns its own cyberinfrastructure in Lebanon. Mobile phones made the Arab Spring possible. Before the Arab Spring, the most successful anti-Mubarak street protests in Cairo either had been organized by bloggers or were about the persecution of bloggers. In times of crisis, troops are sent to defend the one hotel in the port city where the digital switches that connect the country to the global economy are kept cool in the air-conditioning.

While digital media have made it harder for radical ideologies to captivate the imagination of large numbers of people, information technologies themselves have captured the public imagination. Perhaps most surprising is how technology standards themselves have become a civic issue: an important one—an issue that has an impact on how all other policy issues play out.

Most extremist groups never succeed because their ideologies fail to resonate with enough of the people they claim to be fighting for. Social media make it much easier for people to check facts and figures and sources, and to see how the meaning of words and images have been put into the service of political power. Not everyone checks all the facts all the time when radicals try to use digital media. But it takes only a few people to do this and provide the needed corrections and counterclaims.

Fourth Premise: Social Media Solve Collective Action Problems

Ahmed Maher was born in Alexandria, Egypt, in 1980. The next year Hosni Mubarak took the reins of political power, starting a thirty-year run as the country's dictator. This means that Maher had little experience with political alternatives growing up. When two police officers dragged Khaled Said from a cybercafé and beat him to death, Maher and others decided to act.[57] The images of Said's body lying in the morgue—images shared by SMS—were proof of abuse. A few people fed up with violent security services are not enough to drive a revolution. That takes collective action.

Fighting for democracy and freedom presents the mother of all collective-action problems. Why risk tear gas and rubber bullets for an uncertain outcome? Everyone might benefit if you oppose the abuses of a ruling elite. Any one person alone must weigh the daunting costs, risks, and uncertain impact of standing up. What makes an authoritarian government authoritarian is that someone probably will watch you and punish you for your participation.

Too many die at the hands of brutal government-security officials. Increasingly, however, people document the suffering of their loved ones. In Iran, in 2009, Neda Agha-Soltan was shot dead at a street protest, and the video of her blood pooling in the streets of Tehran inspired immense public outrage.[58] This video inflamed the largest protests since the Iranian revolution of 1979. In Tunisia, in December 2010, Mohamed Bouazizi's self-immolation depressed Tunisians, then enraged them to open insurrection.

In Syria, in April 2011, Hamza Ali Al-Khateeb, a thirteen-year-old boy, was brutally tortured and then killed, helping to fuel a civil war.[59] In Bahrain, in August 2011, Ali Jawad al-Sheikh was killed when a police tear-gas canister struck him in the head. These victims focused popular anger. Or, more accurately, their stories were carried by digital media over wide networks of family and friends. The stories made people realize that the risks of individual inaction were greater than the risks of collective action.

Having information technologies that can carefully document a government's failings means that people start to evaluate the costs and benefits of collective action in different ways. If a government terrorizes its people selectively and secretly, most citizens will decide that the risks of rising up in opposition are too great or just too uncertain. But information technologies help people understand what happens if they do nothing. In other words, they start to recognize the costs of staying home, of staying out of the fray: the possibility that random acts of violence might affect them. Taking to the streets no longer seems like risky behavior. Staying at home and doing nothing becomes the real risk.

It's hard for people to figure out when to join a social movement. As Mancur Olson argues in the *Logic of Collective Action*, most groups are doomed to fail for structural reasons.[60] In the abstract, a big group is likely to fail because with so many people in the group, each individual gets only a fractional amount of benefit by contributing to the collective good. A small group is likely to fail because it lacks the resources to have an impact. The best structural scenario for collective action is a big group with lots of shared resources and a few entrepreneurial members willing to do most of the work and figure out the incentives and punishments needed to move everyone else along.

While this is a powerful way of looking at groups in the abstract, it is no longer the best way of explaining why some social movements succeed and others fail. Fundamentally, Olson's way of looking at the world assumed that members would have only occasional contact with each other. Sharing grievances, discussing problems, and acting on solutions would involve only occasional synchronization over broadcast media and through social networks.

When it comes to understanding today's popular uprisings and digital-activism campaigns, we can't forget that communication among group members is usually continuous, if messy. Digital media are almost perfectly aligned with social networks. They are synchronous and two-way. The content is infinitely copy-able and mashable. As Ethan Zuckerman is correct to point out, we often use social media to flock together, strengthening our existing networks.[61] Yet digital cosmopolitanism is driven by both a social problem and the information technologies that coordinate solutions. People will learn, adapt, ask for help, and build community, if the social problem is worth solving.

As Lance Bennett and Alexandra Segerberg argue, fresh thinking is needed to understand how digital media solve collective-action problems.[62] They describe a new logic of "connective action" that explains why digital networks allow collective action in ways that would surprise Olson. Their argument is that digital media help to personalize contentious issues, so that we can all do a better job interpreting the real risks and benefits to participating in collective action. This helps to explain why Maher—and so many others—decided to act in concert against Mubarak. And it helps to explain why spontaneous temporary teams have been able to use voluntary digital mapping projects to solve collective-action problems that have remained intractable for decades.

Looking around the world, research has found that a large proportion of digital activism projects have failed. The vast majority of crowd-sourced maps are started and never used.[63] One of the key findings from global research on digital activism is that bad ideas and poorly executed civic projects fail quickly, while good ideas and effective online projects spread quickly.[64] The overall impact, over several years, has been an impressive list of problems tackled: a growing list of domains in which digital networks have solved collective action problems that, for many years, had not been resolved.

Fifth Premise: Big Data Backs Human Security

If the trends hold steady, by 2020, almost all of humankind will be online. Almost all the books will be there. All the communication between people who aren't in the same room will be digitally mediated, and the physical goods we send to each other will depend on logistics managed by digital systems. Having so many people communicate about so many issues over so many devices creates immense amounts of data. Not all of it is analyzable, and even that which is analyzable is not necessarily useful. Moreover, if the internet of mobile phones and computers produced big data, what will the internet of things generate? For the most part, big data helps people tackle some of the most intractable human security problems.

The internet of things is only going to make big data gargantuan. Much of what we have is often called "dark data." Such data is unstructured, unprocessed, and not easily turned into information, much less wisdom. The hype around using big data to predict flu vectors raised unreasonable expectations.[65] What makes big data important for international affairs is that so much of it is now networked and geolocational. Some seventy-five billion apps have been downloaded for smartphones so far, and there are some two billion smartphones out there. If the average phone has thirty-eight apps on it, and the apps ping a server three times a day, the network generates 226 billion location points.

Think about what you might want to flag in a dataset of credit card purchases. Say you wanted to identify anyone in the United States who researches bomb making online, and then buys bomb-making supplies. By 2020, there will be around 174 billion noncash transactions each year in the United States, including ones for credit cards and paper checks that now are processed electronically.[66] That means 5,500 transactions per second. Each transaction generates several kinds of data about the buyer, the seller, the location, and the product.

To flag a bomb maker with bomb-making supplies, the data miner would have to pull something recognizable from all that data. And if the bomb components are also generating a data trail, then the information needs to be carefully merged. So the number of financial transactions will grow in parallel, presenting a significant analytical challenge. Some people would argue that the government should not be studying such data at all. There are some reasonable trade-offs for protecting democracy. In this example, flagging people who read about bomb making and then buy bomb-making ingredients is reasonable.

Computer scientists have a formal definition of "big data." For them, big data often involves millions of cases, incidents, or files. It involves data generated by many people over many devices. It is measured in terabytes, petabytes, or more. Big data is important to the rest of us for several reasons. We used to have micro-level data about social life—the everyday portraits of interaction generated by journalists and sociologists. We used to have macro-level data, generated through telephone surveys, national censuses, and other large datasets that compared entire countries. Big data, in contrast, provides even richer evidence about particular people and broad social trends, and links it through meso-level data about social ties and your changing behavior over time.

Moreover, we used to categorize data by using our intuition to come up with labels. Now we can let categories emerge from the data through a process we call machine learning. In art, this is the difference between making a sculpture by imposing a design, on one hand, and, on the other, bringing the design out of the rock. Big data is using ever more sophisticated computational models of political behavior. The internet of things is going to make it even easier for researchers and political consultants to use big data in ever more sophisticated computational models to predict political behavior.[67]

In recent years there has been an increasing number of cases demonstrating how big data, often combined with social-media campaigns, can serve to alert both the public and political leaders. Some kinds of monitoring problems are related to arms control, the drug trade, or human trafficking. Many more are related to human security, and connected to health or labor standards that have an impact on the quality of life for significant numbers of people. Big data, when collected purposefully and interpreted well, supports human security broadly. The startup Premise, for example, uses mobile apps to collate pricing information for the world's food staples, from onions to milk.[68]

The internet of things is going to have a big impact on current events. The job of monitoring a problem, verifying that something can be done, and complying with expectations for solving it is tough, especially in global contexts. Fortunately, there is a growing number of vigilante watchers, citizen journalists, hacktivists, and whistle blowers.

In many countries, the government is also the largest employer. And payroll is a big target for corrupt officials. So any system that helps the government pay its employees properly makes the entire economy a little more transparent and efficient. In Afghanistan, when the government started paying its police officers using "mobile money" through mobile phones, many officers were surprised at the size of their paychecks. Some thought they had been given a raise, but it turns out that the new system simply cut out the middlemen who had long been taking their cut.[69] Local bureaucrats could no longer carve out their portion, and funds were suddenly flowing right from the public purse to the public employees. Using digital networks in this big way not only makes it easier to manage the economic role of the state, it improves the personal financial security of public employees.

Even the smallest of NGOs has the capacity to build its own communication campaign and watch either government or corporate behavior. Another reason data is important is that it is useful for catching the lies of government leaders who can't quite admit the truth about bad trends in their country. Everyone knows that China's economic forecasting data is bad. But its carbon emissions data is also bad—making it tougher to study trends in global warming.

Research teams went back over a decade of statistics on carbon emissions that each Chinese province had published. Then they compared the sum of provincial reports to the totals from the national report, and the numbers simply didn't add up. The national-level statistics revealed a 7.5 percent annual increase in emissions. Altogether, the provincial-level statistics added up to an 8.5 percent annual increase—but the numbers should have been the same.

It may seem like a small discrepancy, but we are talking about China. By 2010 this amounted to a difference of some 1.4 billion tons of carbon pollution a year.[70] Were the bureaucrats in Beijing understating the problem, or the bureaucrats in the provincial capitals overstating their figures? The first lesson is that China's economic growth is coming with an environmental cost. The second is that when you let people plumb through data, they find inconsistencies. Sometimes the mistakes are malicious and politically motivated, and sometimes they are not. But spotting numbers that don't add up is an important step in framing and fixing a problem.

Around the world, open-data movements are improving the quality of governance. As we learned in the previous chapter, this is not just about governments but about ways of organizing civic life. It won't matter whether China's government adopts an open-data policy any time soon—the internet of things will generate enough data to keep many China watchers busy.

People are creatively playing with data, some of it gleaned from reluctant governments. Some of that data feeds interesting predictive markets. Such markets don't always work well, and because they trade in odds, you can count the number of times they don't work at all. The models are getting better, and specialists use them to gauge popular and expert opinion on likely political outcomes. When will a dictator fall from power? Will Russia claim more territory from neighboring countries? Someone is taking bets, and someone else is data mining and running experimental conflict models.[71]

With global supplies of data about us being used globally, we are going to have to rethink our assumptions of national sovereignty. States may recognize one another as independent territories. Some states may even be sovereign in that they effectively control the people and resources within their borders. But they will have limited control over the internet of things within their borders.

In fact, people have used the internet aggressively to reassemble some almost forgotten identities, almost always at the expense of national identity. This manipulation has occurred mostly within authoritarian regimes, which seem to see ever more breakaway republics, autonomous zones, and rebel enclaves. Even some democracies have begun to suffer, with increasingly potent claims for cultural autonomy arising. Indeed, almost every democracy has a subculture that has become more aggressive in demanding political independence.

Spain's Catalans, Belgian's Flemish, and Great Britain's Scots have all used the internet to organize bold claims for political self-governance. Australia's aboriginal communities, Norway's Sami and Canada's Innu are groups with something in common: they are territorially disparate national minorities who have used digital media to build their collective identity, lobby for recognition, and improve self-governance. Many don't even live in the lands of their ancestors.

But community members in Sydney, Toronto, and Oslo have reached out to their far-flung family and friends. They have reconnected, and have revived the idea of homelands in Australia, Nunavut in Canada, and Samiland in Norway. In an internet of things that connects so much more of our daily lives with the lives of others, people will probably use the technologies to promote the identities that mean the most to them. For people using the internet to develop a political life, national identities haven't always been the highest priority.

Defining the Pax Technica

The pax technica is a political, economic, and cultural arrangement of social institutions and networked devices in which government and industry are tightly bound in mutual defense pacts, design collaborations, standards setting, and data mining. Over the past quarter-century we have learned a lot about the political impact of new information technologies. Given that our internet is evolving into an internet of connected things, it makes sense to apply what we know in some conservative premises. If these are the premises of the pax technica, what are the consequences—desirable or otherwise?

We have begun an extended period of stability brought about by the dominance of an internet built and maintained by Western democracies. This expansive network of devices has allowed viral social movements that are massive, networked, leaderless, temporary, and multi-issue. Power flows to people and organizations that control digital media or that do creative things with information infrastructure.

Political conflict and competition, in domestic and global contexts, occur over or through information technology. The internet is now standard issue as a weapon for elites seeking social control, and for activists seeking to solve collective-action problems. Conflict and competition are expressed through digital media, from start to finish.

Historically, significant new information infrastructure has ushered in periods of prolonged stability. Reorganizing and rebuilding streets, reconsidering the layout of towns, investing in public transportation and communications systems have all had their payoffs in predictable interactions.

This is part of what explains the rise of Rome and Britain. Rome's roads provided the network that stitched together vast, conquered territories in a way that both defined the empire's economy and demarked the core and periphery. Britain's naval networks defined that empire's economy, and connected people and resources in expansive ties of core and periphery. These empires were not free of violence, but they enjoyed sustained peace that allowed for incredible innovations alongside significant exploitation of conquered cultures.

In the pax technica, the core and the periphery are not territorially assigned but socially and technologically constructed. Or, rather, what connects us is not fixed infrastructure like roads and canals, but pervasive devices with connected sensors. Stability will take the form of cyberdeterrence, new forms of governance, marginalized radicalism, more clans and clubs, and better security for more people.

This doesn't mean that the notions of the core and the periphery are irrelevant. Instead, core and periphery are relevant in terms of culture, status, language, media sophistication, information skills, and social capital. Such attributes may appear geographically distributed. Geography and infrastructure are certainly interdependent. What explains the distribution of social inequalities will increasingly be information access and skills, not physical access and territorial placement.

If there are positive things we can say about the political stability that the internet of things can provide, what are some of the big threats to this stability? What are the best strategies for deepening the encouraging impact of social media over the negative consequences?

5 FIVE CONSEQUENCES OF THE PAX TECHNICA

In a world in which most of our political, economic, and cultural lives are mediated by networked devices, power lies in setting technical standards. Simply put, you either set technical standards or you follow them. International tensions over competing technology standards are only going to increase as governments and firms identify the engineering protocols, licensing arrangements, and telecommunications standards that will allow them to use the internet of things to advance their goals.

Being purposeful about the design of the internet of things is the safest way to export democracy. However, the internet of things is being built over the internet we have now. For each of the premises about the political internet in the previous chapter, there's a consequence for the emerging internet of things.

1. Major governments and firms will hold back on inflicting real damage to rival device networks for fear of suffering consequences themselves—a kind of cyberdeterrence against debilitating attacks.
2. Even more communities will be able to replace their failing governments with institutional arrangements that provide distinct governance goods over the internet of things.

3. The primary fissures of global politics will be among rival device networks and the competing technology standards and media ecosystems that entrench the internet of things.

4. People will use the internet of things for connective action, especially for those crypto-clans organized over networks of trust and reciprocity established by people and mediated by their devices.

5. The great new flows of data from the internet of things will make it much easier for security services to stop crime and terrorism, but unless civil society groups also have access to such data, it will be difficult to know how pervasive censorship and surveillance really is.

Even some of the most banal engineering protocols for how the internet works can have immense implications for political life. If the Russians, Chinese, or Iranians can put those protocols to work for their political projects, they will. Technology standards used to be left to technocrats—the experts who actually understood the internet. But political leaders of all stripes have seen how information technology shapes political outcomes, so they've taken over and exacerbated technology disputes in predictable ways.

Empire of Bits—A Scenario

Imagine that the internet is an empire.[1] It's the year 2020, and the empire is made up of some eight billion people and thirty billion devices. Both constitute the empire, because while people have political values, devices report on both behaviors and attitudes. People have power inasmuch as they are citizens who can use their devices and are aware of how information from those devices is used by others.

The key infrastructure for this empire comprises advanced networks of cables and wireless connections. Being an active citizen of this empire can mean participating in political life through a laptop, mobile phone, or another chipped device. Even those without laptops or mobile phones have an impact because data about their economic, cultural, and social lives feeds political conversations. Their behavior, since they interact with the internet of things, generates data too.

This economic empire has upward of $4 trillion worth of economic exchanges, making it the fifth-largest economy after the United States, China, Japan, and India.[2] The speed of its transactions is mind-boggling, and the pace of its economic growth is rapid. Facebook is one of the empire's biggest provinces, with more than a billion citizens and the device networks that have been given permission to share in Facebook's data streams.

Most people on the planet have varying degrees of membership in this empire—their clout demonstrated by the clarity of data about their attitudes and behavior and the savvy with which they can control the data about themselves. In comparison with other empires, this one is constantly expanding in coverage through the addition of more devices. In the early years much of its population lived in North America and Europe, but now its citizens are spread around the world. In 1989, most internet users were found in the United States, and there were around 900,000 of them. By 2015, around 900,000 new citizens from around the world were joining *each day* and almost all of the 4.3 billion existing internet addresses had been given to devices.

Fortunately, in this scenario, by 2020 a new addressing system will allow every human-made object to have an address.[3]

Like empires throughout history, there are also persistent inequalities. In wealthy parts of the empire the internet is fast, mobile-phone connections are dependable, and the internet of things is transmitting useful information that improves product design and user experience without compromising privacy. This means that even more of the benefits of being connected to the global-information economy accrue in those wealthy neighborhoods. In other places the infrastructure is a little older, and not surprisingly, the possible health and welfare benefits arrive unevenly.

When motivated, this empire rivals the United States and China for political dominance. More important, some countries are determined to stop the internet from having too much political power. They fight back by meddling with its infrastructure, discouraging open standards, and censoring and surveilling its inhabitants. Sometimes they even try to build new internets and subnetworks.

As an empire, the internet of things certainly has its enemies and rivals. The group Reporters Without Borders regularly classifies the "State Enemies of the Internet."[4] Bahrain, China, Iran, Syria, and Vietnam are the governments that most consistently use connected devices to censor and surveil their people and poison other device networks. One of the founding provinces of this empire, the United States, occasionally appears on the list for its government's aggressive surveillance programs.[5]

The Russians, Iranians, Saudis, and Venezuelans try to work alongside the internet empire to various degrees. They still face a digital dilemma when they try to get the economic benefits of participating in the empire without the political risks to their own authoritarian rule. Even the U.S. government gives mixed signals, having built this empire's foundations but with elements of its government doing so much to meddle. Smaller regimes fear the empire. Some try to coexist peacefully with it, some try to undermine it. The Chinese aggressively undermine it by designing rival technologies on parallel networks. They are still trying to build their own internet of connected things.

The latest sensors and mobile-phone technologies connect the inhabitants of the pax technica. This empire is key to the economic life of all the smaller countries, and the innovations of its high-tech industries have an impact on wars around the world, determine the fate of political leaders in many different kinds of governments, and bring communities of interest and identity together.

Its inhabitants have to put up with surveillance—from marketing firms, political lobbyists, academics, and national-security services. And keeping the legal checks and balances over these actors requires constant vigilance.

Moreover, not everything this empire produces is wanted. Every day it generates more and more spam, porn, hate speech, and bad ideas. Sometimes it seems like an empire of anarchy, but that doesn't mean that there aren't occasional leaders, moments of consensus, or deliberate actions. The opinions of its inhabitants matter, and they can make or break an entertainer, a corporation, or a political movement. Temporary teams form, and when they are motivated to act, they can have an impact, whether we are talking about huge parts of our culture, business, or politics.

Of course, the internet isn't quite a country, a superpower, or an empire. Thinking about scenarios for the future is less about predicting specific outcomes than about being prescient about the trends that are likely to continue. We will be layering the internet of things over the internet infrastructure we've built so far, with consequences for each of the five safe premises discussed in Chapter 4. Thinking about the likely political consequences of expanding device networks may help us identify ways to prevent the worst of the possibilities from materializing. What are the likely consequences of expanding device networks for the way we organize, communicate, and do politics?

First Consequence: Networked Devices
and the Stability of Cyberdeterrence
Each major period of political history is defined by its military. Or to put it another way, the major weapons of the time have historically helped to define political realities. The stability of the Cold War rested on nuclear deterrence. In the years ahead, global peace will be maintained by cyberdeterrence, and any balance of power among state actors will be achieved through either cyberwar or the perception of other actors' abilities to wage it.

So many governments and economies will be so dependent on the internet of things that massive attacks using it will be less likely. The proliferation of internet devices will result in a balance of power akin to when a handful of nuclear states held an uncomfortable balance of power. There will be an analogous possibility of accidents, and there will be regional conflicts where minor cyberskirmishes erupt. Based on our recent history, the prospects for all-out cyberwar diminish as the internet of things spreads.

Cyberwar has developed in interesting ways. At first, there were occasional and targeted attacks for specific military secrets. The earliest examples of cyberwarfare date back to the 1980s and 1990s, when Soviet and later Russian hackers went after U.S. military technologies.[6] It wasn't until the turn of the current century that the first cyberwar incidents and campaigns occurred: multiple attacks with retaliations and strategies, involving multiple people who had clearly taken sides or had been sponsored by sides. In this second phase, the attacks were mostly about embarrassing opponents by defacing their websites and demonstrating superior skills.

We've entered a more serious phase in recent years, with extended exchanges of cyberattacks by trained professionals and contracted freelancers. They go after both military and corporate secrets. The public performance of these attacks is important, and getting news coverage for a hack can be a strategic bonus. Many of the more recent attacks are about stealing intellectual property, or about crippling an opponent's ability to develop intellectual property. Cyberwar involves the professional staff of established militaries that no longer just act in response to offline events but are trained to respond to the last cyberattack.

For example, when the United States bombed the Chinese embassy in Belgrade in 1999, Chinese hackers went after several U.S. government websites and brought down the White House website for days. But in 2001, Israeli and Arab hackers went after each other, and each other's internet service providers, for four months. Today, the assault from firms and state agencies in China is constant.

While "cyberpeace" would certainly be preferable to cyber-war, the internet of things will probably support two dependable dynamics of conflict. First, deterrence sets up a kind of stable understanding among all the parties with the skills to damage one another's information infrastructure. Second, one of the most dependable rhythms in current events arises from the cycles of learning and creativity that come with digital media—and how it's used for political ends. Democracy advocates and civil-society leaders try something creative and new. This can throw ruling elites off guard.

When surprised, ruling elites sometimes respond violently and ruthlessly. Sometimes they are so unprepared that they falter, stall, surrender, or make major concessions. Elites in neighboring countries learn from their fallen peers and adapt. Other elites, in neighboring countries, learn. They observe and develop the counterinsurgency strategy that uses device networks for entrapment, propaganda, and the other strategies of social control.

The cascading events of the Arab Spring provide an example of how these cycles work. Digital images of people who suffered at the hands of government security forces circulated among young Tunisians and Egyptians. These images inspired massive turnouts at street protests. Equipped with mobile phones and Facebook, protesters were able to coordinate themselves with an information infrastructure that government officials were not used to manipulating. In frustration, the government eventually tried to disable the country's entire information infrastructure—to pull the plug on its internet access. Elites in Bahrain, Jordan, Morocco, and Saudi Arabia watched closely. When protests emerged in those countries—a few weeks after momentum had built in Tunisia and Egypt—governments were ready with new strategies. Morocco and Jordan offered immediate concessions, and the civil unrest dissipated. Saudi Arabia offered financial concessions—incentives including money for food, rent, and other basics totaling $130 billion to its citizens—while also making a military response.[7] Bahrain, with help from the Saudi government, went straight to a military response, with a heavy-handed crackdown on anyone taking part in a protest.

Moreover, the regimes embroiled in homegrown uprisings found that they could use Facebook and Twitter for their own ends. Because activists were using these digital media to coordinate protests, state security services could use the same tools to coordinate arrests. New users, with false profiles, popped up and called protesters to particular intersections, where those who came were arrested. Police could follow the social networks of protest leaders to map out the ties between organizers and followers. Regime spin doctors could monitor the communications of protest organizers and be ready for questions from journalists. At the beginning of the Arab Spring, Facebook and text-messaging services were tools for the protesters. By the end, Facebook and text messaging were counterinsurgency tools for ruling elites.

Around the world, protesters try tactics with new devices, and regimes learn. Then protesters try new tactics. The civic use of digital media almost always outpaces that of governments. Even governments with some technical capacity often limit their strategies to surveillance and censorship. Every new device that gets connected to the internet gets repurposed in some way. The user, a bot, or a hacker does things with device networks that technology designers never anticipated.

Authoritarian regimes learn slowly, and tend to be reactive. They often have to hire hackers to attack civil-society groups. Other than hiring censorship firms from Silicon Valley, they tend not to be creative with new tools. Activists learn quickly, and they are desperate and creative. Increasingly, they get tech support from civil-society groups, governments, and citizens in the West. The internet of things will be a powerful weapon for the political actors that know how to use it. Perhaps most important, the value and strength of the internet of things will not rely directly on any particular government's stability.

Second Consequence: Governance Through the Internet of Things

Digital governance solutions thrive when established institutions fail and networked devices are available. The important consequence of social-media mapping is that working around scheduled events like elections or humanitarian challenges help community leaders prepare for unscheduled events. Early mapping projects often flopped, but having a group meet and practice coding was useful practice. When projects flopped or had a negligible impact, organizers learned from their failure, such that when some other collective action problem arose— election violence or a sudden forest fire—the social capital and networked devices were available. What we learn from all those social media–mapping projects is that a few altruistic people with even modest technology skills can have a significant humanitarian impact. What do those projects teach us about the prospects for governance over an internet of things?

These days, many experts speak of "governance goods" instead of governance. Governments are supposed to provide goods like working sewage lines and dependable electricity. They are supposed to provide more abstract benefits like trustworthy policies, reasonable banking rules, a postal service, and security from internal and external threats. In times of crises, governments are supposed to provide access to food and shelter. Device networks, when people are encouraged to be creative, can make a wide variety of governance systems more efficient. As discussed in Chapter 4, widespread networks of mobile phones have made food markets more efficient by eliminating waste and reducing wild variations in price. Even more widespread networks, of devices equipped with sensors, can provide more of such stability under the right conditions.

In places where these goods and services aren't provided, people make their own arrangements and fare as best they can. So if there are no trustworthy banks, for example—or no banks that will serve a particular community—mobile phones with banking apps provide an adequate substitute. For us in the West, mobile banking may not seem like an important governance good, because we have a host of stable banking options. In parts of Africa, Latin America, and Southeast Asia, mobile banking systems provide much-needed governance goods.

When states fail, people use digital media to build new organizations and craft their own institutional arrangements. Policy wonks in Washington, D.C., rarely use the term "failed states" anymore, but regardless of the term used, an unfortunate number of governments have ceased to function. Indeed, state failure doesn't always take the form of a catastrophic and complete collapse in government. States can fail at particular moments, like election time, or in particular domains, such as in tax collection. But the mobile phone doesn't take the place of your member of parliament or your government; it substitutes for governance.

For example, in much of sub-Saharan Africa, banking institutions have failed to provide the poor with financial security or the benefits of organized banking. This is due to both a lack of incentive to serve the poor as a customer base and to a regulatory failure on the part of governments that try to establish stable and secure banking regulations for countries. These days, in response, whenever or wherever financial institutions have failed whole communities, mobile phones support complex networks of private lending and community-banking initiatives. Plenty of other large projects involve institutional innovation through technology, so let's evaluate a few.

M-Pesa is a money-transfer system that relies on mobile phones, not banks or the government.[8] Airtime itself has become a kind of transferable asset alternative to government-backed paper currency.[9] M-Pesa is popular in Kenya, but almost every country in Africa has an equivalent service because the banking sectors are either corrupt, too small, or just not interested in serving the poor. Since the "governance good" that can come from having a banking sector that gets some regulatory oversight is missing, people have taken to using their phones to collect and transfer value.

Moreover, they make personal sacrifices to be able to have the technology to participate in this new institutional arrangement. Kenya-based iHUB Research found that people would forgo meat at mealtime if doing so would save enough funds to allow them to make a call or send a text message that might result in some return.[10] In the first half of 2012, M-Pesa moved some $8.6 billion, so this isn't a boutique service.[11] Phone credits are currency that isn't taxed by the government. To put this in some context: a typical day laborer in Kenya might earn a dollar a day, but the value of personal sacrifices for mobile-phone access amounts to 84 cents a week.[12] Two-thirds of Kenyans now send money over the phone.

Politics is about who gets infrastructure, and maps are the highly politicized index of how people and resources are organized. Maps are a key artifact of political power. As discussed in Chapter 3, people in one of Nairobi's uncharted slums made their own digital map specifically for the purpose of identifying public-infrastructure needs and levying their own taxes to help pay for urgent repairs.[13] Ushahidi, the online-mapping platform, can claim many important victories in the battle to provide open records about the demand and supply of social services.

The political power that can come from digital media is the power to let people write and rewrite institutional arrangements. In some parts of the Philippines, the justice system has largely collapsed. So vigilante groups equipped with mobile phones and social-networking applications have organized themselves with their own internal governance system to dispense justice. They deliberate about targets and negotiate about tasks, and they are responsible for upward of ten thousand murders in Manila.

And these days, when individuals feel that their government is not providing the governance goods needed in specific domains, digital media provides the workaround. Average Americans who felt that the U.S. government was not doing enough to support the Green Movement in Iran in 2009 could dedicate their own computational resources to democracy activists. Citizens unhappy with government efforts at overseas development assistance turn to Kickstarter.com to advance their own aid priorities. The next cyberwar might be started by Bulgarian hackers, the Syrian Electronic Army, or Iranian Basiji militias, but it might also be started by Westerners using basic online tools to launch their own Twitter bots.[14]

Even when state failure is partial, or perhaps especially when state failure is partial, people increasingly organize to provide their own governance goods through the internet. For example, when the local government in Monterrey, Mexico, failed to provide public-warning systems about street battles between drug gangs and the military, desperate citizens developed their own public-communication systems. And once in a while there is an example of how governments in wealthy democracies can fail to provide governance at a key moment or in a key domain. During Hurricane Sandy, open-data maps both provided the public with emergency news and information and significantly expanded New York City's capacity to serve citizens in crisis.[15]

Government is not the only source of governance. Technology-led governance is not always a good thing. The internet of things will probably strengthen social cohesion to such a degree that when regular government structures break down, or weaken, they can be repaired or substituted. In other words, people will continue using the internet of things to provide governance when government is absent.

Third Consequence: From a Clash of Civilizations to a Competition Between Device Networks

Information activism is already a global ideological movement, and competition among device networks will replace a clash of civilizations as the primary political fault line of global conflict. Samuel Huntington famously divided the world into nine competing political ideologies, and described these as largely irreconcilable worldviews that were destined to clash.[16] What is more likely, in a world of pervasive sensors and networked devices, is a competition among device networks. The most important clash will be between the people and devices that push for open and interoperable networks and those who work for closed networks.

The dominance of technology over ideology has two stabilizing consequences. The first is that information activism is now a global movement. Every country in the world has some kind of information-freedom campaign that allows for a consistent, global conversation about how different kinds of actors are using and abusing digital media. The second is that the diffusion of digital media is supporting popular movements for democratic accountability. Some Silicon Valley firms build hardware and software for dictators, and as I'll show in the next chapter the serious threat to the pax technica comes from the rival network growing out of China.

Many civil-society groups, even those not concerned with technology policy issues, now think of internet freedoms as human rights. People mobilize themselves on information policy, and civil-society groups have taken up technology standards as civic issues. The reasons are evident: civic leaders realize that their ability to activate the public shapes their political opportunities, and political elites realize that their capacity to rule depends on their control of device networks. The result is that every country in the world has an active tech community that is connected to a global alliance of privacy and information-freedom groups. Some of these activists started their work through the Global Voices network.[17] Others came to technology issues when their websites were attacked, or when their broadband connections got throttled by national ISPs.

In terms of political opinion, they run the spectrum, and many are more interested in fast-streaming access to content about the Eurovision contest or distant soccer games than in political news. Some are libertarian, others progressive, some conservative, and some a mix of all three. But they are all often dedicated to pushing back on onerous government regulations over their internet access, and some participate in, or eagerly read about, technology issues from the Electronic Frontier Foundation and the Center for Democracy and Technology, tweet about the latest reports on their country from the Open Society Foundation or Reporters Without Borders, run Tor Project software quietly on their home equipment, and even participate in training sessions from the Tactical Technology Collective.

Even the most banal technology standards in the poorest of countries get scrutinized by civic groups emboldened by John Perry Barlow's "Declaration of the Independence of Cyberspace," or Hillary Clinton's arguments that internet freedoms are a foreign policy priority.[18] When I visited Tajikistan a few years ago, the government simply didn't have an employee who was in charge of public-spectrum allocation. The Aga Khan Foundation was providing a staff of three Western-educated computer science interns to help set policy.[19] Who would have thought that young civil-society actors would want to weigh in on how the public spectrum gets allocated, or want to attend specialized International Telecommunication Union meetings on internet protocols?

Digital media have allowed civil society to bloom, even in the toughest of regimes. Telecommix, Anonymous, and CANVAS are committed to teaching civic leaders to be more tech savvy.[20] These groups include Nawaat in Tunisia and Piggipedia in Egypt.[21] Piggipedia may not bring many prosecutions of police torturers. But it breaks the fear barrier for citizens, helps victims find a way to respond, and reminds police that they should be accountable public actors. It used to be that the nuances of internet protocols were left to engineers concerned with system efficiency and business opportunities. A growing number of people, especially in the West, have a basic literacy about cookies, privacy, and censorship. This is a good thing.

Information activism has developed a powerful ideology of its own, one that can shape the spending priorities of governments. A cadre of civic groups has real clout in technology policy. At the end of 2012, activists' lobbying helped to prevent the ITU in Dubai from giving governments the power to interfere with the internet. In the United States, they defeated the antipiracy legislation known as the Stop Online Piracy Act because it went too far in protecting the interests of record labels and media companies. They took on the Anti-Counterfeiting Trade Agreement in Europe and worked aggressively on Brazil's bill of internet rights, the Marco Civil. They have successfully campaigned against national firewalls in countries like Pakistan and aggressive cybercrime laws in the Philippines. Sympathizers and concerned citizens contribute the computing power of their home machines to Tor networks, particularly in times of crises for democracy movements in other countries.

As Muzammil Hussain demonstrates, the pace of collaboration between the state department and Silicon Valley quickened after the Arab Spring.[22] In its aftermath, information activism has grown more sophisticated, and moved into a transnational environment, as demonstrated by Western democracy–initiated stakeholder gatherings. AccessNow, the main organization that lobbied corporations to keep communications networks running and pressured technology companies to stop selling software tools to dictators, organized the Silicon Valley Human Rights Conference in November 2011.[23] The event was sponsored by Google, Facebook, Yahoo!, AT&T, Skype, and other technology firms, and it brought together the corporate leaders and foreign policy officials of major Western democratic nations to design policies for corporate social responsibility in the interest of international human rights. Similarly, the governments of the United States, the Netherlands, Sweden, and the European Union all created formal funding programs totaling more than $100 million to support digital activists working from within repressive regimes.

At least seven conventions and conferences have been brokered by the foreign policy offices of key Western democratic countries since the Arab Spring. These meetings have brought together information activists and technology corporations. U.S. Secretary of State Hillary Clinton has referred to this interesting mix of brokered meetings as "twenty-first-century statecraft." Information activists increasingly find themselves working diplomatically with Western foreign policy makers on one hand and also targeting and lobbying businesses to stop building technologies for repressive regimes.

In several European countries, the civic groups that formed under the banner of defending "internet freedoms" have successfully become political parties. The Pirate Bay, a file-sharing website, inspired widespread interest in intellectual-property issues. Enough interest, in fact, that small new political parties have popped up around the world, taken the name of the Pirate Party, and fielded candidates for elected office. By 2015, Pirate Parties had started in more than forty countries, and dozens of party members had been elected to city and regional governments in countries around the world.[24] In short, technology access, digital-cultural production, and information access have become civic issues. Civil-society groups from across the political spectrum are now concerned with privacy and information policy because these things have such an impact on their work.

Even outside Europe, networks of like-minded technology advocates turned their online activism into Pirate Parties.[25] Significant numbers of voters have put Pirates into office, raising the visibility of technology-related issues and improving the public's literacy on intellectual property law reform, public-spectrum allocation, and telecommunications standards setting. In some authoritarian regimes, where governments worked for decades to close down policy domains from open debate, discussion of technology policy was tolerated. In part, this was because many authoritarian states either had no capacity to set internet policies or did not think them worth controlling. This made technology access a civic issue that was safe to mobilize on—at first. With newfound skills in organizing, educating, and lobbying governments, these public-interest groups have been able to expand to other issue domains.

People sometimes say that the internet doesn't "cause" democracy. Or "it's the people, not the mobile phones." But people and their technology are often impossible to separate. Try to imagine your life without your mobile phone or your internet connection. Or try to tell the story of the Arab Spring, the Occupy movement, or any recent international social movement without mentioning digital media. You'll find yourself with an incomplete story. Many of the people involved with these movements are eager to talk about the devices and media that are their tools of resistance. Their technology and their story go together.

Political scientists have found similar causal narratives when they compare many different kinds of political changes over time: media use, as a causal factor conjoined with others, often provides the best explanation for political outcomes. In other words, economic wealth, social inequality, and education are robust predictors of democracy on their own. But their explanatory power grows when these variables are paired with media use. Social development is important, but understanding diffusion patterns is even more important. Predictors of spatial proximity, networks, and digital media have become among the most important parts of any democratization analysis. In some ways, the relationship between media exposure and democratization is pretty straightforward. Exposure to news and information from democracies raises the hopes of those living in authoritarian regimes. The introduction of device networks means more knowledge about the standard of living in Western democracies and the rise of incentives to create sustainable democracy at home.

Of course, device networks get embedded in cultures, introducing different patterns of adoption and local variations in political values. The causal relationship is there. In fact, research suggests that it was media control that prevented democratic norms from spreading around the Middle East, and the introduction of digital media and social media that undermined these same controls. In the years leading up to the Arab Spring, whole cohorts of young people across North Africa were developing political identities under the very noses of aging elites who had ruled for decades. This was possible because these young people used their devices to build their own trust networks. Young people, and democracy advocates, will continue to do this with new device networks. A good many of these will be distributed networks in which people may not have met but have validated one another through personal ties and trusted cryptography.

Fourth Consequence: Connective Action and Crypto Clans

The fourth consequence of the internet of things is that connective action will solve more and more collective problems. Whereas Ahmed Maher had a powerful reason for joining a popular uprising for democracy, Eliot Higgins's interest in the Syrian civil war is tough to figure out. As the blogger "Brown Moses,"

Higgins develops his hobby for a cause. He tracks the weapons that appear in images and video coming out of conflict zones. He records who is holding which assault rifle, where armaments appear to be stored, and what impact those weapons have.

His research has been used by media outlets and politicians. He culls his observations from hundreds of YouTube videos filed by journalists and citizens caught up in the conflict. When the Syrian government denied using cluster bombs on its people, he had the video evidence to catch its lie. He found barrel bombs when the Russians said none were being used. He has counted and catalogued shoulder-launched, heat-seeking missiles and found Croatian weapons that must have come to the Syrian opposition with Western assistance. The systematic evidence about the regime's use of ballistic missiles, which he collected at his home computer, triggered an Amnesty International investigation.

Getting good information from countries in crisis is always difficult, and it is even more difficult when ruling elites have a lot to hide. Higgins isn't Syrian, hasn't been to Syria, and has no long-term ties to Syria. He has no weapons training and has taught himself to recognize weaponry using online sources.[26] Why would he get involved?

Brown Moses is only one of many people and projects that have put the interests of a few or one to work for the many. Indeed, as more people get mobile phones and smartphones, more digital-activism projects have appeared. The main antidote to dirty networks of gun runners is the attention of social media. Modern political life is rife with examples of how people have used social media to catch dictators off guard and engage their neighbors with political questions.

Ushahidi, the mapping platform for crowd-sourced knowledge, has a good record of problem solving. It may be one of the largest and most high profile of such providers of connective action, but it's not the only one. Uchaguzi is a platform built specifically for monitoring the Kenyan election in 2013.[27] In neighboring Nigeria, researchers find that the number and location of electoral fraud reports is highly correlated with voter turnout.[28] This means that social media are starting to generate statistically valid snapshots of what's happening on the ground—even in countries as chaotic as Nigeria. Still, connective action doesn't just happen through crowd-sourced maps.

Indian Kanoon, an online, searchable database on Indian law, has opened up a whole swath of data to the average person.[29] Many Indians are proud of living in the largest democracy in the world, but it is difficult for average citizens to understand Indian law. The text of an act can be extensive, and finding the small section of law that has a bearing on any specific issue can be difficult. Extracting the applicable sections from hundreds of pages of law documents is too daunting for nonlawyers. Moreover, laws are often vague, and one needs to see how they have been interpreted by judges. In Indian law libraries, the laws and judgments are often maintained separately, making it difficult for average citizens to link relevant laws with judgments and precedents. Indian Kanoon is helping to make the law more accessible.

In Russia, Liza Alert helps coordinate the search for missing children.[30] Other sites track complaints about poor public services and coordinate volunteers.[31] People use the internet to track corruption at universities in Kenya and Uganda.[32] India's

I Paid a Bribe project inspired similar projects in Pakistan and Kenya.[33] Altogether, these projects don't simply make up a scattered network of do-gooders. Over the past decade we've seen civil-society actors develop digital tools for engaging with the public and with public policy makers. India's Kiirti platform relies on the public to identify problems, crowd-sources the process of verification, involves a civic group in identifying relevant policy makers, and summarizes the trends for policy makers.[34] Today, digital activists are often found at the center of new social movements.

Vladimir Putin's hold on national broadcast media has been so tight that civil-society actors turned to the internet, and there they bloomed—and found solidarity. In an important way, the Russian internet became the home of the effective opposition, because it was there that the best investigative journalism, anti-corruption campaigns, and groups like Pussy Riot found audiences. In Tunisia and Egypt, before the Arab Spring, the largest civic protests were either organized by bloggers or were about the arrest of bloggers.

People have used device networks to produce very different kinds of social movements. During the revolutions of 1848, the civil unrest of 1968, and the popular uprisings of 1989, formal hierarchical organizations drove political events. They were well organized, had clear leadership structures, and were armed with ideological or nationalist zeal. They often had the savvy to put then-new media to work for them and their propaganda: leaflets, radio, and cassette tapes carried messages to supporters. Social networks were important for binding together people within the same working class, and the result was large cohorts of like-minded socialists, nationalists, and freedom fighters who acted in concert and were quick to form political parties.

Today's social movements are distinct. They are much less about class and race, and no longer so driven by prepackaged ideologies and high-profile, charismatic leaders. Now social movements are temporary networks of networks, sharing grievances and a negotiated action plan. The first wave of protesters on the streets of Tunis and Cairo were young and middle class—they weren't just Islamists, Marxists, or the poor from urban slums. Street protests involved networked groups, many of which had formed around specific digital initiatives. They governed themselves in peculiar ways.

The modern social movement is a temporary team of linked, smaller networks with a shared memory of past interactions with state police and the expectation of future contact with one another. This peculiar aspect of network politics is also what can make such social movements brittle. They may act swiftly and impressively and massively, but they may not last. As we've seen from the Occupy Movement, the Arab Spring, and many major national protests since, networked social movements have a difficult time becoming political parties. They have a tough time staying in the long game.

It would be impossible for all of us to form opinions on all issues, much less volunteer to help solve every political problem. The key reason social media are helping people solve collective-action problems is that they link up the people who *do* have the ideas and energy to work together. In a sense, digital networks help organize knowledge and address ignorance. And this happens even in authoritarian regimes, where ruling elites see the young as uncontrolled and the young see themselves as power-less. Social media give power to the rationally ignorant.

Social media certainly helped Ahmed Maher to find his network in Cairo, and they allowed Eliot Higgins to make his contribution to international affairs. Both formed digital clubs of peers with shared interests, willing to act together. Both have generated useful knowledge and information for the larger clans of activists, journalists, and interested publics.

Yet the impact of social media extends beyond information supplies and personally compelling calls to action. Social media encourage collective action precisely because information is embedded in a social context. It's what we do with the information, how we act in our daily lives, that ultimately helps us address the challenges of our era. These days, there are two qualities that are unique about the connective action enabled by digital media: we maintain clans, and we join clubs. These two terms from anthropology and economics have usefully specific meanings for the internet of things.

Facebook facilitates the formation of clans and supports clan identity. Essentially, it is a service that allows clans to stay connected on a daily—even hourly—basis, and across international borders. And perhaps not surprisingly, these clan ties, digitally maintained, are behind resurgent subnationalisms around the world. Political identities that had been effectively subsumed for decades have surged back because their communities preserved cultural knowledge, and various diasporas reconnected, sent money, and carried political issues around the world.

A crypto-clan is a group based on actual or perceived kinship and descent that we actively maintain through new information technologies. Your extended family, your close friends on your block, your high school pals: these can all be the basis of your clan affiliations. Race, ethnicity, gender, and sexual orientation are among the many sources of identity we have. But groups we actively maintain with Facebook and Twitter and crypto-clans, and they are important because they are elective. We don't always choose our family ties, but we do choose how to maintain those ties digitally. We don't know everyone in our crypto-clans, but within a few steps it is easy to get to know the other people in the clan. Broad norms of trust and reciprocity permeate crypto-clans, and our clan networks are great sources of information. A crypto-clan is a collection of weak ties, mediated by device networks, that we get to curate for ourselves.

In contrast, digital clubs are the smaller groups of people whom we know more directly. Trust and reciprocity aren't the only criteria for membership. Active engagement is the norm, and providing collective good is the goal. Like crypto-clans, digital clubs are elective. So they are important because as groups they have even more of our loyalties and goodwill—they are collections of strong ties, digitally mediated. We use digital clans to gather information, and we join digital clubs to change the world.

In an authoritarian regime, it's risky to attend a small protest. You'll get beaten up and arrested, and the regime will learn who you are. The internet of things will transform the way people participate in politics by making their devices agents of information about opinion and behavior. Researchers found that during the Arab Spring in Cairo, one-third of protesters were protesting for the first time.[35] The vast majority of people who attended the first day learned about the event online, and they all used their devices to do some kind of documentary work to discover what was going on. These weren't individuals attending a small protest, they were a clan of like-minded citizens linked by shared grievances and values.

Exactly who controls the process by which networked devices generate politically valuable information remains to be seen. Will it be the users who buy them, the companies that make them, or the governments that surveil networks? Currently industry and government have the most systematic means of rendering politically meaningful information from networks of devices. Unless civil-society groups fight for a radical change in the understanding of how the internet of things is to be used, the "balance of power" favors those who have the cryptographic skills to use and manipulate the networked devices around them.

In many authoritarian regimes, there is a disjunction between ruling elites, who see the young as dangerous and shiftless, and the young, who see themselves as unable to act. If there is something to be learned from the peculiar social movements of recent years, it is that young people who feel disenfranchised will teach themselves technology tricks. This does not mean that the world will be full of cryptography experts. Those who are skilled in cryptography will be able to form clans and clubs, using basic security tools with their devices to create distributed networks of trust.

Fifth Consequence: Connective Security and Quality of Life
The fifth consequence of the internet of things is that people will look for more and more ways to use device networks to improve human security. Along with connective action, there is some

security that comes from having infrastructure that can be studied for big-picture trends and small deviations. We might start to think of this as "connective security," namely, the improvements in our ability to track good trends, monitor bad behavior, and make reasoned security decisions with high-quality data. First, the data generated by the internet of things is likely to improve our ability to watch our governments and construct new governance mechanisms when established ones fail. Second, all this data will improve our ability to follow and extinguish dirty networks of criminals, drug lords, and political strongmen.

Big data refers to the kind of information you can get from lots of people using lots of different technologies. Mobile phones, video-game consoles, email accounts, website log files, and a host of other appurtenances of consumer electronics generate immense amounts of information about people's interests. Even more valuable is information about people's behavior that comes from analyzing big data. Through big data a country can address its own health-and-welfare problems, and direct smart military operations such as the one that killed Osama bin Laden. Just as social media can be used to coordinate democracy advocates or as a tool for social control, big data can be used for solving social problems, surveillance, or manipulation.

Big governments often "frame" problems as being manageable. In countries where new media are dominated by the government, journalists sometimes just pass the official perspective on to citizens. The U.S. government framed the Hurricane Katrina disaster as a crisis it was addressing efficiently.[36] The Russian government claimed that it had a handle on the forest fires of 2010, a claim refuted by community maps. The Japanese government claimed that the radiation leaking out of the Fukushima Daiichi nuclear plant was under control. Mapping by several people, including web designer Haiyan Zhang, proved otherwise.[37] Social media allow alternative frames, and sometimes they allow communities to frame problems, victims, and culprits before a government can spin the facts.

Along with tracking the bad behavior of governments, big data makes it possible to track international networks of criminal activity. Digital media have made it much easier to coordinate different networks of crime fighters and get them working in concert across international borders. To combat dirty networks requires input from different kinds of organizations: local police, intelligence agencies, militaries, media outlets, academics, NGOs, and, perhaps most important, average citizens.

One of the most pernicious networks includes the ties that form between dictators, drug lords, gun runners, holy thugs, and rogue generals. But police also form networks, and they have better resources for putting together big data for themselves. The rising drug violence in South America, for example, has motivated police across several borders to work out joint security plans. Experienced officers from Colombia and Chile now help those in Nicaragua and Honduras.

Indeed, the fact that mafias use information technology often makes it easier to map out their networks. Most anticrime initiatives are national. Almost every country in the world has seen a boom in local anticorruption initiatives over social media. Some are organized by average citizens putting together the pieces to solve mysteries, and some come from major technology firms targeting drug cartels.[38]

Dirty networks are effective because the connections between powerful criminals are tenuous and closely guarded. The nodes in such networks often have chosen to hide in Latin American or African jungles because the roads that link their compounds to the rest of the world are not easily watched. The Lord's Resistance Army has been able to move through the Congo, South Sudan, and the Central African Republic for decades. A single road, passable only during the dry season, weaves through the western part of South Sudan. But tracking technologies improve faster than these thugs' ability to discover new hiding spots.

In some cases, people are able to muster significant information resources toward undermining the credibility of poor leaders, then organize opponents for a coordinated push. These information cascades start as small examples of citizen journalism, efforts to document police abuse, or political jokes. They can grow to topple a dictator. Dirty networks are not always governments, but bad governments are often networked with drug lords, corrupt generals, or holy thugs. Social media can be used against those kinds of political actors too, as a way of undermining their control or as a way of coping during moments of extreme violence.

A powerful country is going to be one that has the capacity to use big data to solve its own domestic social inequalities. Such a country can use the internet of things to outmaneuver enemies, and use social media to deepen cultural relationships even when government leaders are hostile with one another. The powerful country is going to be the one that can intelligently analyze a continuous flow of information from neighbors, friends, and enemies.

During the Cold War, security analysts, pundits, and scholars studied patterns of alliances and camps. Any country that didn't fall neatly into a camp raised doubts about its strategic intentions. Now the perception of camps is less relevant, and indeed it is possible for a savvy international player to use digital media to create the impression that it is an active member of several camps.

Big data is useful not just for understanding the global connections between political actors. It can be useful as well for understanding small and local sociotechnical systems. For example, Sandy Pentland argues, in *Social Physics*, that an immense amount of organizational complexity can be captured with what he calls reality-based research involving sensors and log files.[39] Yet big data analysis certainly has critics, and the big data debate is relevant for thinking through the impact of the internet of things.[40] The internet of things is going to make big data truly gargantuan. Successfully analyzing the data that can be collected by a world of networked devices will itself be an engineering challenge. Unfortunately, the hyperbole and enthusiasm for the "social physics" of big data analysis has kindled new excitement about data mining. Most people, most of the time, value their privacy online.[41] Big data analysis can be good for modeling behavior, but is unable to reveal peoples' attitudes and aspirations.

The Downside of Connective Security

Immense privacy violations will arise if we aren't careful about the ways we use big data. The potential for abuse—by politicians or businesses—is significant. Ron Deibert documents this well in *Black Code*.[42] And large amounts of data, smartly analyzed, have been solving a great range of problems. Incompetence and bad policy cannot trump good information forever.

The prospects for big data analysis are exciting because of the possibility of exposing previously unsuspected causal relationships, and confirming the ones we've found too tough to demonstrate in other ways. For example, demand for white vinegar soared in Guangzhou, China, in 2002.[43] SARS was spreading, but the central government was denying any public-health emergency. There was a public perception that vinegar would help clean kitchens of whatever bacteria was spreading. It took outside analysts to connect the trends and identify the public-health crisis. Knowing something about how the public deals with problems can help alert a government of the need for action.

During the SARS epidemic, Chinese authorities tried to censor messages that mentioned the disease. But people figured out how to talk in ways that couldn't be quickly detected by authorities. So the Chinese spent significantly more money to beef up their surveillance. Even now, they don't seem to be able to control everything. They seemed less able to control topics than mobilization efforts. They've had to reduce censorship efforts, limiting themselves to framing and discouraging physical logistics among protesters. And the Party seems unsuccessful even in this effort.

Information is always prone to political interpretation, and there is a proven capacity for bad analysis in this process. Experts, at least, know that correlations are spurious. The best example of this is the entertaining analysis produced by some economists revealing that the rate of production of butter in Bangladesh was a good correlate of the rise and fall of the S&P 500 stock index between 1981 and 1993.[44] Big data can still be put to work for the social good, but all data needs to be interpreted and contextualized.

And, of course, bad governments can be good at big data analysis. For example, the *Miami Herald* has a good record of breaking important news stories about Latin American politics. In March 2012 it reported that sensitive data about Venezuelans was being kept in Cuba.[45] Government databases, voting records, citizenship and intelligence records, and more were being stored in server farms outside Havana. To an outsider, it might be silly to think of Havana as a more secure city than Caracas. What makes this interesting is not that data about a country's citizens was being transported out of country. The Russian mafia has bought the credit card records of many U.S. citizens. Data-mining firms in Texas maintain detailed profiles of Argentina's citizens, and there is a global trade in data about people from all corners of the world.[46]

What is important in the Cuba-Venezuela connection is that the government did not choose to house important information with a firm or in a place that has good security or stable infrastructure. Data warehouses across the United States and Europe have such features. The network ties between Venezuela and Cuba are so strong that they overcame any technical logic to file storage. The data did not simply need to be stored, it needed to be stored with political compatriots that would share the same expectations of surveillance and social control. So while bad analysis of big data is a real danger, there is also a proven capacity for big data analysis by bad governments.

It will be challenging to extend and adjust our expectations for privacy and definitions for bad behavior. Criminals will want to manipulate the internet of things as much as anyone. If drug smugglers can figure out how to hack the cranes working in the Belgian shipping port of Antwerp to move their illicit cargo around, they will try to make use of the equipment in our homes too.[47]

If there is anything we can learn from the evolution of the political internet, it is that there are persistent threats to its openness. For some communities, the internet of things will bring collective insecurity: bad analysis can come from big data; privacy violations appear to be the norm for big data collection; and bad governments are capable of doing big data analysis too. In the end, however, social problems are easier to solve with good data. At the very least, the internet of things will greatly expand the network of devices capable of carrying political information. So what are the greatest threats to the greatest of these networks? What rival interests and actors have the ability to weaken the pax technica or limit it in some way?

6 NETWORK COMPETITION AND
THE CHALLENGES AHEAD

The internet of things will help bring structure to global politics, but we must work for a structure we want. This is a challenging project, but if we don't take it on our political lives will become fully structured by algorithms we don't understand, data flows we don't manage, and political elites who manipulate us through technology. Since the internet of things is a massive network of people and devices, structural threats will come from competing networks. There are two rival networks that seriously threaten the pax technica: the Chinese internet and the closed, content-driven networks that undermine political equality from within the pax technica.

The Chinese internet is already the most expensive and elaborate system ever built for suppressing political expression. The Chinese are trying to extend it by exporting their technologies to authoritarian regimes in Asia and Africa. Russia, Iran, and a few other governments are also developing competing network infrastructures. The Chinese government controls the entire network, the network is bounded in surprising ways, and the network can, and does, mobilize to attack other networks.

Within democracies, the real threat comes from campaigns against net neutrality and for new limited-access networks. When technology and content companies make it easier to get certain kinds of content, they create subnetworks of social and cultural capital. Putting research ideas behind paywalls—especially research that has been publicly funded—also creates competing networks of information. Political parties love to keep the party faithful in bounded information networks. So while our main rivals are external, there are also internal threats to the strength of our information networks. Information technology is only a means to a political end. However, the internet of things will be the most important means.

My Girlfriend Went Shopping . . . in China

The Chinese have a deliberate and vigorous strategy for combating the empire of the pax technica. When China's political elites met in 2013 for a national congress to choose their leaders, propaganda from Beijing dominated official news feeds. Yet over social media, it was a boyfriend's complaints about his girlfriend that went viral.[1]

> Whenever my girlfriend goes shopping, she tends to get overly serious and way more than just fidgety about the whole thing. It always interferes with my usual pace of life. Anyway, she calls the shots at home, so [I] can't complain. As my girlfriend stipulates, when it approaches her shopping date, I can only make working plans for up to three days, and if I go on a business trip, I need to get her approval first. These past few days I've been sitting on pins and needles, praying to God that I don't do anything wrong to ruin her good shopping mood.

. . . I guess as long as she buys things for me, I shouldn't complain too much. . . .

She usually doesn't pay attention to me when she shops. Well, you do your shopping, and I'll tend to my own business, I think to myself. So I take out my phone to surf the net a bit. But before I can open even one page, she pops up immediately: "You can't just get online like this when I shop! What emails are you checking? If you dare check one more, I'll deactivate your Gmail account!"

Many of China's digerati are so used to censorship and surveillance that they quickly learn to talk in complex metaphors and trade tricks for getting access to the tools they want to use. In this coded rant, a young student on Renren—the Chinese Facebook—compared the 18th Chinese Communist Party Congress to his girlfriend's latest shopping trip. The party is a vain girlfriend who can be jealous and abusive but is sometimes generous and occasionally responsive. The girlfriend watches everything he does online.

For a long time, China watchers worried about the daunting size of China's active army. It wasn't always well trained or well equipped, but it was disciplined. It was the world's largest. These days there are more than 2.5 million men and women in the country's active military, another 800,000 reservists, and 1.5 million paramilitary members. It's a military built to project power outside the country and maintain control inside the country. These days, battalions of censors provide more social stability than the military muscle of the Chinese armed forces.

The Communist Party has developed a dedicated army to resist the spread of the technologies and values of the West.

By one estimate there are more than 2 million "public-opinion experts," a new category for jobs that involve watching other people's emails, search requests, and other digital output. In other words, the army of censors is as large as the military, and often military units are given censorship and surveillance tasks.[2]

Government agencies need censors, but the government also makes tech startups and large media conglomerates hire their own censors to help with the task of watching the traffic. Pundits have referred to these people as China's fifty-cent army because some get paid small amounts of money to generate pro-regime messages online. But that moniker makes the army of censors seem like freelancers who are inexpensive to hire. In actuality, they are a well-financed force deeply embedded in the country's technology industry.

With demand for these jobs high and stable employment all but assured, young people have to pay for the education to become public-opinion analysts. According to China expert Guobin Yang, organizations like the *People's Daily* can charge up to four thousand renminbi ($650) for four days of training to become an analyst.[3]

Lots of other authoritarian regimes employ censors, but let's put the numbers in perspective. If there are 2 million people occupied with Chinese censorship tasks, and 500 million users, that's one surveillance expert for every 250 people. Aside from the human resources put into censorship and surveillance, China's device networks have three unique features: the government controls the entire network, the network is bounded in surprising ways, and the network attacks other networks.

First, the Chinese government owns and controls all the physical access routes to the internet. People and businesses can rent bandwidth only from state-owned enterprises. Four major governmental entities operate the "backbone" of the Chinese internet, and several large mobile-phone joint ventures between the government and Chinese-owned media giants offer additional connectivity.

An important part of Chinese network control is the way the party controls the intermediaries who build hardware and provide connection services.[4] When representatives of the telecommunications firm Huawei were called before the U.S. Congress to answer questions about the security of the hardware equipment it sells, the company argued that its internal documents were "state secrets."[5] This admission might frighten us, but disclosures by Edward Snowden and from within our sociotechnical system reveal that firms and governments are also tightly bound up in background deals, mutually convenient understandings, and shared norms.

The research on China's censorship efforts finds that the government works hard to support Chinese content and communication networks that it can surveil, and discourages its citizens from using the information infrastructure of the West. In one study, researchers went through the process of launching a social media startup in China.[6] They took notes each time they encountered a new regulatory hoop to jump through, and they kept track of the amount of information they were making available to the state-security apparatus.

They had to hire consultants to help with compliance. They had to open log files and use specific hosting services within

China. Most important, they learned that Chinese censors were not interested in censoring critical blog posts and essays. They were after posts about organizing face-to-face meetings, or messages about the logistics of protesting. The academics collected data on traffic spikes on Sina Weibo, China's equivalent to Twitter. The traffic spikes on controversial topics around which people didn't seem to be organizing themselves were often left to fizzle out on their own. But traffic spikes about organizing protests were quickly shut down—censors want to block connective action. In fact, China's strategy of blocking independent political organization before it can start demonstrates that connective action is what the Party fears most. The government knows that the internet can catalyze activist organizations. So it has designed its internet to be watchable and works to make sure every unsanctioned civic group is stillborn.

Second, the Chinese resist Western device networks by making sure that connections within China are extensive and reliable, and connections to the rest of the world less so. Chinese device networks are bounded by the Great Firewall of China, as we call it in the West, though the more poetic translation is the Golden Shield Project. Some consider it the largest national security project in the world, and its singular task is to protect Chinese internet users from access and exposure to outside content.

By one estimate it cost the Chinese government $800 million to build the Golden Shield. Protecting itself from the Western internet is an ongoing venture.[7] So the Chinese government has a simple two-part strategy for insulating itself from the pax technica. It blocks some technologies and copies other ones. Social-media applications are almost always redesigned for domestic use, and foreign firms operating in China must locate their servers in the country, where state security services can get at them easily. Platform standards and expectations are different from those of the pax technica. Requirements for real-name registration, for example, make the immense task of monitoring message traffic on the internet easier.

Practically, this means that messages from someone in China to someone outside China travel slowly. But the reason they travel slowly is important: the messages must go through one of several key digital servers. In network parlance these servers are mandatory points of passage: nodes that handle all the traffic flowing into and out of China. Such nodes can slow down traffic either because their capacity is limited or because the nodes are inspecting the packets as they go by.

This is deliberate, as the Chinese want to create their own rival internet. As one Chinese technology official said, "The big question is not whether or not China can build a world-class society while fighting the internet, the question is whether or not it can do so while building a giant intranet that is China-specific."[8]

For individual users the impact may be felt in terms of extra lag time and the insecurity of knowing that this delay is because censors are actively searching content. Staring at the screen waiting for an email to arrive may seem like a minor inconvenience we all must suffer at some point. In the United States, we know that marketing firms, the NSA, and internet-service providers also scan and copy content we individually produce and consume. The metadata about our production and consumption is also captured for analysis.

But for China as a whole, the national surveillance strategy means a slow network that gets slower at politically sensitive moments.[9] Imagine if the NSA's spying was somehow even more extensive, as well as completely authorized, unrestricted, and immune to organized protest. Within China, the average internet speed is about three megabytes per second. By comparison, internet connections average three times faster in the United States and the United Kingdom. The bandwidth for traffic flowing from China to the rest of the world is far less than its already low national average.

This means that the government has created a structured way of allowing the Chinese internet to grow, while slowing the passage of content from outside the national network. Extensive monitoring and close collaboration between government and industry allow the Party to preserve the Chinese internet. These factors also mean that the internet of things will struggle to evolve in China. Either the government has to work out ever more sophisticated techniques for monitoring the traffic among devices, or it will have to give up on the idea of surveilling the entire network.

Third, the Chinese government is aggressively assaulting international information infrastructure. Corporate cyberespionage, design emulation, patent acquisition, and technology export are the key weapons of this attack. Some of these are subtle defenses that smack of cultural protectionism, while others are aggressive strategies for attacking outside networks. Cyberattacks on Western news media regularly originate from within China.[10]

The Chinese are trying to win over other nations, co-opting the device networks of poor countries into nodes of their rival network. The Chinese are not just seeking to protect their citizens from the West, they are aggressively expanding their networks to rival the pax technica through cultural content, news production, hardware, software, telecommunications standards, and information policy. In terms of content and news, their rival strategy involves:

- Direct Chinese government aid to friendly governments in the form of radio transmitters and financing for national satellites built by Chinese firms.
- Provision of content and technologies to allies and potential allies that are often cash strapped.
- Memoranda of understanding on the sharing of news, particularly across Southeast Asia.
- Training programs and expenses-paid trips to China for journalists.
- A significant, possibly multibillion-dollar, expansion of the People's Republic of China's (PRC's) own media on the world stage, primarily through the Xinhua news agency, satellite and internet TV channels controlled by Xinhua, and state-run television services.[11]

Whether this propaganda and surveillance system is sustainable with the great volume of device networks to come is a big question. Simply put, China is aggressively building the main rival network. Many kinds of attacks—whether corporate espionage, military provocation, or political manipulation—get launched from China. The Chinese internet is the primary rival to the pax technica.

Despite China's best efforts, there are signs that this block-and-copy strategy doesn't always work. Tealeafnation.com regularly translates and publishes "bad speech" from the country's dissenters.[12] The 2012 policy of requiring new Sina Weibo users to register with their real names has proven tough to enforce, and verifying the identity of the millions of existing accounts almost impossible. Can China continue to keep its citizens on and within its own bounded internet? Other countries want to build their own national internets using Chinese technology, even if it means effectively joining China's network by becoming dependent on China for innovation and by providing backdoor access to Chinese security services. The question remains: will China's rival network actually expand in the years ahead?

While the government of China has worked hard to protect its citizens from the global internet, social media have made it easier and easier for technology users to pry into the private lives of the country's corrupt officials. When the *New York Times* published an investigation into the extensive family assets of China's premier Wen Jiabao, the website was blocked to Chinese users. The government tried to block every effort to discuss the investigation on Sina Weibo, the Chinese Twitter, with only modest success.[13] Only a few months earlier, Bloomberg News had conducted an investigation of a prominent Chinese politician, Xi Jinping.[14] Xi was rising to the position of Party secretary general, but his family wealth had been accumulated in suspicious ways. Bloomberg's websites were blocked in China, but inside China the conversation continued on Sina Weibo.

There are ever more examples of how Chinese citizens use social media to push the limits of free speech. Li Chengpeng was a sports reporter for several decades. After the 2008 Sichuan earthquake killed more than 80,000 people and exposed the limits of the government's ability to help people in crisis, he started writing about his nation's social ills. He found his voice on Sina Weibo, and eventually more than 6.7 million users found him. His investigations on corruption now reach an immense audience, and he can bypass the state-controlled media. More important, he teaches his audience about which issues to track and cautions them to be savvy.

It is not possible to connect to the global internet "a little bit." China has done much to shield its citizens, but people in the West can now have more exposure to Chinese culture and politics than ever before. This, too, comes with security implications. Essentially, it means that official agencies like the NSA can spy on the communications between Chinese officials. It also means that internet users around the world can investigate trends in China.

In 2011, after that same earthquake in Sichuan, Georgetown professor Phillip Karber noticed that the hills in the affected region had collapsed in strange ways. China was sending radiation experts to the disaster zone. So he started investigating with a team of undergraduate students. After three years of work, the investigation exposed a network of underground tunnels used by China's Second Artillery Corps. The students translated thousands of pages of documents, studied Google Earth, scanned Chinese blogs, read military journals, and groomed their own contacts in China for information, producing a revised estimate of the number of nuclear weapons operated by that country's military. Their work was the largest body of public knowledge yet published on China's nuclear arsenal.

Experts in the United States have been estimating that the Chinese nuclear arsenal is relatively small, consisting of between eighty and four hundred warheads. But investigations found that the tunnel network was designed to support up to three thousand warheads. Professional analysts were skeptical.[15] Partly as a result of public exposure, Chinese officials began revealing more information about the network of tunnels. In the United States, the report sparked a congressional hearing and renewed conversations among top officials in the Pentagon.

While the internet can help the West learn about life in China, it is also a conduit for public diplomacy for both sides. The Voice of America (VOA), an official U.S. government media organization, has a viral online video show called OMG! Meiyu, in which Jessica Beinecke, in fluent Mandarin, presents Western pop culture and discusses Western slang.[16] Her two-minute videos have a broad audience. Perhaps most important, however, her followers interact with her and with one another. They ask her questions, and she responds in subsequent web videos. It's not the traditional broadcast model long used by the Voice of America; rather, it's an online exchange through and about current cultural phenomena. Explaining what the "Final Four" is or what it means to "get stuffed" is not high-level diplomacy.[17] It is a crucial part of the VOA's mission. Since China's elites are finding it harder and harder to reach their young people over broadcast media, cultural outreach like this from the West over social media is even more important.

Technology designers and users in the United States and Europe may have done much of the initial creative design work to bring the internet into being. Silicon Valley still designs the tools that people in many countries use—especially those in the pax technica. But China is redesigning such tools for its citizens.

Increasingly, it manufactures the hardware that many interests depend on.

Still, building a controlled, bounded, and invasive network limits the ability of the system to do what the internet has been good at, namely helping people solve collective action problems and generating big data for human security. Crowd sourcing and altruism flourish in open societies with open device networks. In closed societies, such projects generally appear only in times of crisis, when people see that building a temporary, issue-specific governance mechanism is worth risking the wrath of a tough regime. This means that tough regimes rarely can call on their publics to contribute altruistically to a government initiative.

Research on China's feeble attempts at open government demonstrates that crowd sourcing doesn't work well, and in China's context is better thought of as "cadre sourcing."[18] This is because the kinds of information sought by the government have already been distorted by the government, enthusiastic cadre participants are more likely to report favorable information than accurate information, and news about independent crowd-sourcing initiatives don't circulate far. Only during complex humanitarian disasters do people decide to take on the risks of contributing quality information. As device networks spread, civic initiatives will always have more positive impacts in open societies. Authoritarian societies are structurally prevented from making use of people's goodwill and altruism. In the bounded device networks of an authoritarian regime, crowd-sourcing initiatives are likely to create negative feedback loops and big data efforts are likely to generate misinformation about the actual conditions of public life.

With this pernicious structural flaw, how much faith should we have that China's rival information infrastructure will stay rivalrous? What will the government have to do to retain control of the internet of things that evolves within its borders? Or to return to the metaphor, what do you do about that "jealous girlfriend" who gets ever more domineering during her shopping trips?

> I've known her for such a long time, from the first time we went shopping together to this eighteenth time. There have been sweet moments, but there were also moments of despair. She once tortured me [horribly] and made my life worse than death. She also took it upon herself to take care of me when I met with natural disasters. Despite all these headaches she's been giving me, she has made some progress over the years nonetheless. She still has many shortcomings, but she's more and more open to my criticism now.

Authoritarian, but Social

China's aggressive efforts to build a rival network are not the only form of resistance to the device networks of the pax technica. The Russians have been successfully pioneering another strategy, one emulated by Venezuela, Iran, and some of the Gulf States. The Russian gambit is not to build its own network from the ground up, it is to join the internet by sponsoring pro-regime internet users to generate supportive commentary online. One of the ways that they do this is through summer camp.

Every summer, thousands of young Russians gather at camps around the country. They do what everyone does at summer camp: make friends, flirt, get into a little trouble, breathe some fresh air, and exercise. These Russian summer camps, however, are different from those found in the West. For many decades these have been state-organized affairs, and have involved indoctrination into national myths. Teenagers learn a few survival techniques, meet new people from other parts of the country, and listen to patriotic lectures. But in the past few years, Vladimir Putin's updated summer camps have also involved social-media training.

Putin's political base was not online, so he enlisted the program directors of summer camps around the country to begin teaching social-media skills. Compared with the Chinese government, the Kremlin was not quite able to dedicate the same resources to content production, surveillance, and censorship. By 2006 tens of thousands of teenagers at youth camps around the country were getting short courses on Putin's vision for the country and training sessions on blogging.

The most notorious of the camps is called the Seliger Youth Forum, and it has been held outside of Moscow each summer since 2005.[19] The camps are organized by a pro-Kremlin youth movement called Nashi ("Who if not us?" in Russian). This group taps into the long tradition of summer camps for Russian youth.

Putin has his favorite summer camps, but he has also promoted particular universities and colleges over others. I visited Sholokhov University in Moscow in the summer of 2012, shortly before that year's Seliger Forum. The university was organizing a large international conference on social-media analytics. To the Western participants it quickly became clear that our hosts didn't want to study the development of social movements online. Their questions were different from ours. We had sociological research questions, but they had very practical questions about identifying the members of social networks.

During a tour of the campus, we learned that the university had begun as a home for humanities departments. But as the more prominent Moscow State University fell out of favor, Sholokhov began receiving more state funding. These days, it has the county's top-ranked undergraduate degree in lie-detector equipment operation and interpretation.

Despite devoting increased attention to recruiting youth, Putin's regime found that while it could dominate broadcast media, civil-society groups were flourishing through social media. When Russia's civil-society groups and political opposition were squeezed out of the national broadcast media, marginalized in parliament, and pushed off the streets, they went online. Discussion forums, websites, and blogs establish the sociotechnical system needed to sustain their ties during Putin's political freeze.

Civil-society leaders found a large, sympathetic audience of other disaffected Russians online. Russian internet users—even now—tend to be young, live in cities, and have had some experience with life overseas. Compared with the country's new users, Russia's established internet users are slightly more educated, slightly more liberal, and slightly less interested in Putin's nationalist visions for the country. The opposition had been marginalized, but its creative use of the internet actually allowed it to flourish.

Whereas broadcast media are useful for authoritarian governments, citizens use social media to monitor their governments.[20] For example, in early 2012, rumors circulated that a young ultranationalist, Alexander Bosykh, was going to be appointed to run a Multinational Youth Policy Commission. A famous picture of Bosykh disciplining a free-speech advocate was dug up and widely circulated among Russian-language blogs and news sites, killing his prospects for the job (though not ending his career).[21]

These are not simply information wars between political elites and persecuted democracy activists. The organization and values of broadcast media are very different from those of social media. Putin is media savvy, but his skills are in broadcast media. The Kremlin knows how to manage broadcast media. Broadcasters know where their funding comes from, and they know what happens if they become too critical. Indeed, Putin's changes to the country's media laws are specifically designed to protect broadcast media and burden social media.[22] In Russia, critics have been driven into social media, where they have cultivated new forms of antigovernment, civic-minded opposition. Russian political life is now replete with examples of online civic projects achieving goals the state had to give up on.

One of the recent battles over network infrastructure within Russia involved the government's webcam system for monitoring elections. To prepare for the election, the government spent half a billion dollars on webcams for every polling station in the country. With widespread skepticism about the transparency of Putin's regime, this move was designed to improve the credibility of the electoral process.

The election was held on March 4, 2012, and within a few days a highlight reel of election antics went up on YouTube.[23] For Putin's political opponents, the webcams demonstrated citizen indifference toward the election. For Russia's ultranationalists, the webcams revealed no systematic fraud. Alas, the elections commission decided that video evidence of fraud was not admissible. The video feeds were not systematically reviewed, and the electoral outcomes were never in doubt. When Russian-affiliated troops rolled into the Crimea in 2014, they did so with the full backing of a coordinated social-media campaign. The cohort of young Russians raised to be active on Twitter and blogs provided strong and consistent messaging on Putin's behalf.[24]

When authoritarian governments try to exploit social media, they rarely have clear and consistent success at the game. There are four strategies for a dictator who seeks to "go social." Governments can pay people to generate pro-government content. Governments can physically attack information infrastructure, or the cybercafés and homes of people who use their internet access for politics. They can use digital media for surveillance, and simply monitor the flow of communication for content that should be blocked or people who should be arrested. Or they can have their security services hack the devices used by civil-society groups.

Putin's Kremlin has developed a comprehensive counterpropaganda response to social media, and this strategic package has bounded Russian information networks. Moreover, the strategies have proven transportable. Now many governments groom, hire, or train pro-government commentators for social-media work. Venezuela's Ministry of Science and Technology has staff dedicated to hacking activist accounts so as to use those accounts to seed dissent and confusion among political challengers.[25] Some governments track the physical location of computers, servers, and other hardware so that such infrastructure can be destroyed if needed. Tech firms in Europe and the United States produce some of the best surveillance and censorship software and hardware available today. Tools like Finfisher, which can remotely activate webcams, are widely available.[26]

Russia Today helps generate content for Russia's social-media networks, and CCTV does the same for Chinese social networks. And just as the Chinese are exporting their network by expanding their infrastructure and training network engineers in other countries, Russia's strategy is being transported. Azerbaijan, Iran, Venezuela, and Cuba now all have cohorts of paid, pro-regime social-media contributors. The success of Russia's digital-media strategy is best measured by a simple outcome—Putin handily retains control and seems as strong as ever at home through a veneer of authorized power.

Other political parties and governments have begun attacking civil-society groups through devices. The Chinese regularly go after Tibetan groups online in the hope of finding activists based in China.[27] Outside of repressive regimes, who would attack a civic group?

In the summer of 2012, WikiLeaks had to beat off a string of hacker assaults.[28] The year before, Amnesty International faced a coordinated attack.[29] Indian Kanoon, the online project described in the previous chapter that is dedicated to making Indian law searchable and accessible, was hit by a denial of service attack in October 2013.[30] Many such attacks have occurred, many of them unreported, and those that do get reported tend to involve large nongovernmental organizations that expose the bad behavior of unethical businesses and authoritarian governments. Coupled with the growth of false grassroots, or "astroturf," campaigns launched by political consultants and lobbyists worldwide, civil-society groups face security challenges and competition for public trust.

Unfortunately, this is the precursor of worse to come. Civil-society groups are largely unprepared for today's cyberattacks, much less the volume of attacks and types of vulnerability that will come over the internet of things. They depend on open-source software, whose performance and security can be uneven, and on free services that include product-placement advertising. They tend to be run by volunteers and strapped for cash; rarely do they have the resources to invest in good information infrastructure. The world's authoritarian governments are better positioned than civil society groups for the internet of things.

Bots and Simulations

Any device network we build will create some kind of what Eli Pariser calls a filter bubble around us.[31] We will be choosing which devices to connect, and those devices will both collect information about us and provide information to us. But the danger is not so much that our information supplies may be constrained by the devices we purposefully select. It is the danger that our information supplies may be manipulated by people and scripts we don't know about.

The word "botnet" comes from combining "robot" with "network," and it describes a collection of programs that communicate across multiple devices to perform some task. The tasks can be simple and annoying, like generating spam. The tasks can be aggressive and malicious, like choking off exchange points or launching denial-of-service attacks. Not all are developed to advance political causes. Some seem to have been developed for fun or to support criminal enterprises, but all share the property of deploying messages and replicating themselves.[32] There are two types of bots: legitimate and malicious. Legitimate bots, like the Carna Bot, which gave us our first real census of device networks, generate a large amount of benign tweets that deliver news or update feeds. Malicious bots, on the other hand, spread spam by delivering appealing text content with the link-directed malicious content.

Botnets are created for many reasons: spam, DDoS attacks, theft of confidential information, click fraud, cybersabotage, and cyberwarfare.[33] Many governments have been strengthening their cyberwarfare capabilities for both defensive and offensive purposes. In addition, political actors and governments worldwide have begun using bots to manipulate public opinion, choke off debate, and muddy political issues.

Social bots are particularly prevalent on Twitter. They are computer-generated programs that post, tweet, or message of their own accord. Often bot profiles lack basic account information such as screen names or profile pictures. Such accounts have become known as "Twitter eggs" because the default profile picture on the social-media site is of an egg.[34] While social-media users get access from front-end websites, bots get access to such websites directly through a mainline, code-to-code connection, mainly through the site's wide-open application programming interface (API), posting and parsing information in real time. Bots are versatile, cheap to produce, and ever evolving. "These bots," argues Rob Dubbin, "whose DNA can be written in almost any modern programming language, live on cloud servers, which never go dark and grow cheaper by day."[35] Unscrupulous internet users now deploy bots beyond mundane commercial tasks like spamming or scraping sites like eBay for bargains. Bots are the primary applications used in carrying out distributed denial-of-service and virus attacks, email harvesting, and content theft.

The use of political bots varies across regime types. In 2014, some colleagues and I collected information on a handful of high-profile cases of bot usage and found that political bots tend to be used for distinct purposes during three events: elections, political scandals, and national security crises. The function of bots during these situations extends from the nefarious case of demobilizing political opposition followers to the relatively innocuous task of padding political candidates' social-media "follower" lists. Bots are also used to drown out oppositional or marginal voices, halt protest, and relay "astroturf" messages of false governmental support. Political actors use them in general attempts to manipulate and sway public opinion.

The Syrian Electronic Army (SEA) is a hacker network that supports the Syrian government. The group developed a botnet that generates pro-regime content with the aim of flooding the Syrian revolution hashtags (e.g., #Syria, #Hama, #Daraa) and overwhelming the pro-revolution discussion on Twitter and other social-media portals.[36] As the Syrian blogger Anas Qtiesh writes, "These accounts were believed to be manned by Syrian Mokhabarat (intelligence) agents with poor command of both written Arabic and English, and an endless arsenal of bite and insults."[37] Differing forms of bot-generated computational propaganda have been deployed in dozens of countries.[38] Current contemporary political crises in Thailand and Turkey, as well as the ongoing situation in Ukraine, are giving rise to computational propaganda. Politicians in those lands have been using bots to torment their opponents, muddle political conversations, and misdirect debate. We need political leaders to pledge not to use bots, but the internet of things will make them easier to use.

Table 3 reveals that bot usage is often associated with either elections or national security crises. These may be the two most sensitive moments for political actors where the potential stigma of being caught manipulating public opinion is not as serious as the threat of having public opinion turn the wrong way. While botnets have been actively tracked for several years, their use in political campaigning, crisis management, and counterinsurgency is relatively new.[39] Moreover, from the users' perspective it is increasingly difficult to distinguish between content that is generated by a fully automated script, by a human, or by a combination of the two.[40]

In a recent Canadian election, one-fifth of the Twitter followers of the country's leading political figures were bots. Even presidential candidate Mitt Romney had a bot problem, though it's not clear whether exaggerating the number of Twitter followers he had was a deliberate strategy or an attempt by outsiders to make him look bad. We know that authoritarian governments in Azerbaijan, Russia, and Venezuela use bots. The governments of Bahrain, Syria, and Iran have used bots as part

Table 3 **Political Bot Usage, by Country**

Country	Year	Polity	Deployer, Assignment
Australia	2013	10	Political parties—hiring bots to promote candidate profile and policy ideas
Azerbaijan	2012	−8	Government—attack opposition, manipulate public opinion about public affairs
Bahrain	2011	−8	Government—attack opposition, manipulate public opinion about public affairs
Canada	2010	10	Political candidates and parties—buying followers on social media
China	2012	−8	Government—disrupt social movements, attack protest coverage, manipulate public opinion about public affairs
Iran	2011	−6	Government—attack opposition, manipulate public opinion about public affairs
Israel	2012	10	Government, military—information war with Hamas and PLO
Mexico	2011	8	Political parties—misinformation during presidential election
Morocco	2011	−6	Government—attack opposition, manipulate public opinion about public affairs
Russia	2011	4	Government—attack opposition, disrupt protest coverage, manipulate public opinion about public affairs, influence international opinion on Crimea
Saudi Arabia	2013	−10	Government—attack opposition, manipulate public opinion about public affairs
South Korea	2012	8	Government—use social media to praise elected head of government
Syria	2011	−8	Government—attack opposition, manipulate public opinion about civil war, misinformation for international audiences
Thailand	2014	7	Government—support coup
Turkey	2014	10	Candidates—give impression of popularity; government—manipulate domestic public opinion

United Kingdom	2012	10	Candidate—give impression of popularity
United Kingdom	2014	10	Government—manipulate public opinion overseas.
United States	2011	10	Candidate—give impression of popularity; National Security Agency—manipulate public opinion overseas
Venezuela	2012	2	Government—attack opposition, manipulate public opinion about public affairs

of their counterinsurgency strategies. The Chinese government uses bots to shape public opinion around the world and at home, especially on sensitive topics like the status of Tibet.

Bots are becoming increasingly prevalent. And social media are becoming increasingly important sources of political news and information, especially for young people and for people living in countries where the main journalistic outlets are marginalized, politically roped to a ruling regime, or just deficient. Sophisticated technology users can sometimes spot a bot, but the best bots can be quite successful at poisoning a political conversation. Would political campaign managers in a democracy like the United States actually use bots?

These days, campaign managers consider interfering with a rival's contact system an aggressive campaign strategy. That's because one of the most statistically significant predictors of voter turnout is a successful phone contact from a party supporter the night before the election. That reality is what has driven up invasive robocalls. In 2006, automated calling banks reached two-thirds of voters, and by 2008 robocalls were the favored outreach tool for both Democrats and Republicans. Incapacitating your opponents' information infrastructure in the hours before an election has become part of the game, though there have been a few criminal convictions of party officials caught working with hackers to attack call centers, political websites, and campaign headquarters. Republican National Committee official James Tobin was sentenced to ten months in prison for hiring hackers to attack Democratic Party phone banks on Election Day in 2002. Partisans continue to regularly launch denial-of-service attacks; attackers consistently target Affordable Care Act ("Obamacare") websites, for example. If an aggressive move with technology provides some competitive advantage, some campaign manager will try it.

So the question is not whether political parties in democracies will start using bots on one another—and us—but when. They could be unleashed at a strategic moment in the campaign cycle, or let loose by a lobbyist targeting key districts at a sensitive juncture for a piece of legislation. Bots could have immense implications for a political outcome. Bots are a kind of nuclear option for political conversation. They might influence popular opinion, and they are certainly bad for the public sphere.

For countries that hold elections, bots have become a new, serious, and decidedly internal threat to democracy. Most of the other democracies where bots have been used have tight advertising restrictions on political campaigns and well-enunciated spending laws. In the United States, political campaigning is an aggressive, big-money game, where even candidates in local and precinct races may think that manipulating public opinion with social media is a cost-effective campaign strategy.

Political campaigning is not a sport for the weak. Neither Twitter nor Facebook is the best place for complex political conversations—few people change their minds after reading comment threads on news websites. Citizens do use social media to share political news, humor, and information with their networks of family and friends. Meaningful political exchanges over social media are prevalent, especially when elections are on the horizon.

There may come a time when the average citizen can distinguish between human and autogenerated content. Of course, with an industry of political consultants eager to have the most effective bots at election time, it is more likely that the public may just come to expect interaction with bots.[41] But for now, with most internet users barely able to manage their cookies, the possibility that bots will have significant power to interfere with public opinion during politically sensitive moments is very real.

Bots will be used in regimes of all types in the years ahead. Bots threaten our networks in two ways. First, they slow down the information infrastructure itself. With the power to replicate themselves, they can quickly clutter a hashtagged conversation, slow down a server, and eat up bandwidth. The second and more pernicious problem is that they can pass as real people in our own social networks. It is already hard to protect your own online identity and verify someone else's. Political conversations don't need further subterfuge.

Badly designed bots produce the typos and spelling mistakes that reveal authorship by someone who does not quite speak the native language. Well-designed bots blend into a conversation well. They may even use socially accepted spelling mistakes, or slang words that aren't in a dictionary but are in the community idiom. By blending in, they become a form of negative campaigning.[42] Indeed, bot strategies are similar to "push polling," an insidious form of negative campaigning that disguises an attempt to persuade as an opinion poll in an effort to affect elections. The American Association of Public Opinion Researchers has a well-crafted statement about why push polls are bad for political discourse, and many of the complaints about negative campaigning apply to bots as well.

Bots work by abusing the trust people have in the information sources in their networks. It can be very difficult to differentiate feedback from a legitimate friend and autogenerated content. Just as with push polls, there are some ways to identify a bot:

- One or only a few posts are made, all about a single issue.
- The posts are all strongly negative, effusively positive, or obviously irrelevant.
- It's difficult to find links and photos of real people or organizations behind the posting.
- No answers, or evasive answers, are given in response to questions about the post or source.
- The exact wording of the post comes from several accounts, all of which appear to have thousands of followers.

The fact that a post has negative information or is an ad hominem attack does not mean it was generated by a bot. Politicians, lobbyists, and civic groups regularly engage in rabble-rousing over social media. They don't always stay "on message," even when it means losing credibility in a discussion.

Nonetheless, bots have influence precisely because they generate a voice, and one that appears to be interactive. Many users clearly identify themselves in some way online, though they may not always identify their political sympathies in obvious ways.

Most people interact with several other people on several issues. The interaction involves questions, jokes, and retorts, not simply parroting the same message over and over again. Pacing is revealing: a legitimate user can't generate a message every few seconds for ten minutes.

Botnets generating content over a popular social network abuse the public's trust. They gain user attention under false pretenses by taking advantage of the goodwill people have toward the vibrant social life on the network.

When disguised as people, bots propagate negative messages that may seem to come from friends, family, or people in your crypto-clan. Bots distort issues or push negative images of political candidates in order to influence public opinion. They go beyond the ethical boundaries of political polling by bombarding voters with distorted or even false statements in an effort to manufacture negative attitudes. By definition, political actors do advocacy and canvassing of some kind or other. But this should not be misrepresented to the public as engagement and conversation. Bots are this century's version of push polling, and may be even worse for society.

DRM for the Material World?

Social media bots are not the only automated scripts working on our digital networks. Most devices, if they are designed to be hooked up to a network, are designed to report data back to designers, manufacturers, and third-party analysts.

For example, when Malaysian Airlines Flight 370 was lost in March 2014, it emerged that some of the best available data on the status of the plane was not the black box recorder designed specifically to preserve data. Nor was it the data from satellites. It was data from the embedded chips in parts of the engine that could be read as evidence that the plane was flying after pilots stopped communicating. In fact, the company that built the plane and the company that built the plane's engines each had its own active device network.[43] Boeing had equipped the plane with a continuous data monitoring system, which transmits data automatically, and Rolls-Royce collected data on engine status. It turned out that Malaysian Airlines was subscribing only to the engine analytics service from Rolls-Royce, but both equipment manufacturers collect immense amounts of data about the things they make long after those things are sold. So the engine parts came with embedded data systems that remained accessible to manufacturers even if the owner of the material devices didn't want to pay for access to the data.

Such data supplies are useful—to many people. They can help manufacturers improve the quality of their products. They can help suppliers understand their consumers. It is not just specialized, high-performance engine parts that now come embedded with chips and networking capabilities. Most consumer electronics, if they download some kind of firmware, software, or content, upload some kind of location, status, and usage data.

One television owner in the United Kingdom discovered this when he started playing around with the settings on his LG Smart TV. Many of us aren't interested in exploring the inner workings of our devices, but those who do sometimes find that data is being sent in surprising ways. When Jason Huntley, who blogs as DoctorBeet, tried to understand what information his new LG TV was collecting, he was surprised at what was being aggregated.[44] The TV could report the contents of any files read from a memory stick. It was reporting what was being watched to LG's servers. The company had to admit that even when users turned this feature off, the device continued to transmit information.[45] The company claimed to need this information to help tailor ads. Huntley discovered that his TV was essentially a computer, and that LG was interested in a long-term relationship with him, through the data about his media interests.

Other manufacturers were quickly caught up in the public debate. Samsung revealed that its televisions were collecting immense amounts of information, and that it has the ability to activate microphones and cameras on its latest models. An investigative team looked at software vulnerabilities and advised owners against giving networked TVs a view of their beds.[46] While device manufacturers might understand the data trail that a device leaves about its use and its users, consumers may not be told of the trail.

As the internet of things expands, increasing numbers of manufacturers will be presented with the opportunity to make money from long-term services associated with what they are making. At the very least they will be presented with the opportunity to collect vast amounts of data that might be valuable to a third party. Experience suggests that most will succumb to the temptation, unless explicit public policy guidelines require manufacturers to make good decisions and help consumers understand what their device networks are up to.

Even more dangerous is the prospect that device manufacturers will begin to use digital rights management (DRM) to protect the families of devices and streams of data they can collect. If the future value of a device is in the data it returns—behavioral data would be valuable to analysts, product data would be valuable to designers—then manufacturers have an incentive to protect this value stream. A precedent for this kind of behavior already exists, as several different industries have pushed for digital rights management to aggressively protect intellectual property.

DRM so far has been mostly used to protect against copyright infringement, particularly of music, video, and other cultural products.[47] When industry lawyers found it difficult to go after individual infringers, they went after internet-service providers who supply the infrastructural support for infringement. Would DRM become an option for manufacturers eager to protect uniquely designed material objects and the data they render? It is hard to know where to draw the line, but the line needs to be drawn by civil-society groups and consumers, not by corporations.

Other Challenges (That Are Lesser Challenges)

A handful of other challenges threaten the stability of the pax technica. Most of them are not structural threats because the information networks that set up problems also become the sources of resolution and security. Criminals and extremists will always find ways to use and abuse the internet of things. One such problem is that new technologies leak across markets and jurisdictions, resulting in political advantages and disadvantages for different actors. It is almost impossible to fully bound a national internet without allowing some connections to the outside world. New technological innovations spread along social and digital networks. So the latest software for encrypting messages passes from activist to activist over Tor servers. At the same time, the latest snooping hardware ends up in the hands of repressive regimes.

For example, when Bolivia wanted a new national biometric card system, it turned to the Cuban company Datys.[48] Data on Venezuelan citizens is also stored and analyzed in Cuba. In Colombia, as noted previously, equipment provided by the United States for the war on drugs was also used by government intelligence officers to spy on journalists and opposition politicians.[49] Some governments don't confine themselves to spying on their opponents in country, however. The Ethiopian government watches even those political opponents who are part of the cultural diaspora living in other countries.[50]

Blue Coat Systems manufactures equipment to secure digital data, but the same equipment can restrict internet access for political reasons, and monitor and record private communications. When the Citizen Lab's researchers went looking to see where such devices were popping up on public networks, they spotted them as far afield as Syria, Sudan, and Iran—countries supposedly already subject to U.S. sanctions.[51]

But it is not just that people learn about new software from friends, family, and colleagues, or that countries sharing political perspectives also share technologies. In China's case, its information architects go to countries like Zimbabwe to train that country's government in ways to build a network that might serve economic interests without taking political risks.

Drug lords also make use of digital media. Indeed, one of the features of the pax technica is that information infrastructure has helped make different kinds of political actors structurally equivalent. In Latin America, the large drug gangs Los Zetas and MS-13 occasionally cooperate, and frequently battle. In their own territories each rules more effectively than the government, and each competes with the government for the control of people and resources. They conduct information operations by hacking state computers, attacking journalists, and going after citizens who tweet about street shootouts.

Along with competing infrastructure networks, dirty networks will remain the great threat to peace and stability when they've been able to adapt device networks for their goals. Corrupt political families make everyone angry because they usually set about enriching themselves. In the case of Tunisia, as Lisa Anderson argues, the now-deposed dictator had taken family corruption to new levels of effectiveness.[52]

> Ben Ali's family was also unusually personalist and predatory in its corruption. As the whistleblower website WikiLeaks recently revealed, the U.S. ambassador to Tunisia reported in 2006 that more than half of Tunisia's commercial elites were personally related to Ben Ali through his three adult children, seven siblings and second wife's ten brothers and sisters. This network became known in Tunisia as "the Family."

If corrupt ruling families, radicals, and extremists can use digital media to build their ranks, they will try.

The English Defense League, a group that inspired Norway gunman Anders Behring Breivik on his murderous trek in 2011, had only begun two years earlier. It grew from fifty members to more than ten thousand supporters in two years, and the far right group's leaders expressly credited Facebook as its key organizational tool.[53] There are thousands of forums for extremist groups of all kinds—including ISIS—and many have multimedia sites offering streaming lectures, mobilization guides, and social-networking services.[54] Feiz Muhammad, who also inspired the Boston bombers Dzhokhar and Tamerlan Tsarnaev, has generated a great volume of YouTube sermons from the safety of his home in Sydney, Australia. A great volume of white-supremacist videos can also be found online.

At the same time, the internet allows these kinds of groups to seem deceptively powerful. In fact, it is clear that they have less overall impact on public life than the more moderate groups that meaningfully engage with political processes. Certainly some individuals, such as the Tsarnaev brothers in Boston, find inspiration from texts and groups online. They remain, for the most part, pariahs. Most of the purposive organizational activity online is still aimed at furthering the greater good, not at singling out people to hate.

For someone running, or planning to run, a terrorist organization, the internet is a dream technology. Access is cheap and the reach is international. There are tricks for making your online activities anonymous, such as the steganography of embedding messages in photos. Though there's plenty of evidence that such tricks don't fool the very security agencies that have been able to build surveillance tools right into their networks.[55] There can be no doubt that the same technology that enables democracy advocates to convert followers can also let terrorists do the same. Anticorruption organizations use information technologies to track and analyze patterns in government spending, and governments use the same technologies to track and analyze citizen behavior. But we can't forget that all technology users are embedded in communities, meaning that they face social pressures and, more important, they face socialization.

For several years in China, anti-cnn.com collected material from thousands of people who wanted to report on the supposedly pro-Tibetan bias of Western media. The site was launched in 2005 and quickly earned such clout that it was able to become a key node in Chinese cybernationalism, with the power to organize street protests and consumer boycotts.[56] Organizing hatred—or love—on the internet is easy because consumer-grade electronics make it possible for those with just a little tech savvy and a small budget to aggregate and editorialize content. Often this work doesn't involve generating new ideas, it involves only linking up to other people and getting them to provide text, photos, and videos. Putting vitriol online doesn't mean it will have wide appeal.

The Downside of Up

Connective security and connective action have definite downsides. The same information infrastructure that allows friends and family to trade emails allows Russian mafia members to buy the credit card records of U.S. citizens, or terrorists to plan and launch attacks. Governments use the web to advance their territorial claims, interpret (and sometimes forget) history in flattering ways, justify human-rights abuses, or assert regional power. They spy on their citizens. The risk of keeping our online infrastructure open is that some people will use this system for evil. Someone will always try to set up rival infrastructures for social control rather than creativity.

But one of the challenges, when it comes to adding it all up, is that there is no way to know whether the bonds of friendship between people divided by distance and culture are more numerous—or stronger—than the ties that have been destroyed or weakened by digitally mediated communication. Put more simply, are there more good relationships and projects coming out of digitally mediated social networks than bad?

The myth that the internet is radicalizing our society, fragmenting our communities, or polarizing our political conversations makes for a good editorial or news-feature story. It remains more of a news peg than a demonstrable, widespread phenomenon. It may have become easier to find the text of *Mein Kampf* or other tomes of hatred online, but there is much more content and social interaction that has none of that hatred. There are many more mainstream political parties, regular newspapers, and middle-of-the-road political conversations. A balanced look finds examples of how digital media have been useful for both constructive and destructive political engagement. If the internet of things greatly expands the digital network of our political lives, will the network as a whole be more conducive to hate speech?

Attempts by extremists and criminals to use device networks for their hateful and nefarious projects represent a small risk when weighed against the benefits and affordances of the internet of things: if personal data is managed responsibly and civil-society groups can actively participate in the standards-setting conversations, that is. Over the past twenty-five years of internet interregnum, the violent extremists who organized online became easy to identify and catch as a result. The cure—widespread surveillance—may or may not be worse than the disease. Extensive surveillance might put a pall on the mood of the majority of internet users, but the ongoing NSA surveillance scandal seems not to have affected user attitudes. Knowing that the NSA can surveil the Western web and that the Chinese can surveil their telecommunications infrastructure has not had consequences for the enthusiasm of most new users.

With the right conditions, a radical website can galvanize a community of hatred, give individuals a target for their vitriol, and help them organize their attacks, both on and offline. Fortunately, it's rare to get the right conditions for this heady mix of nasty conditions.

Rival Devices on Competing Networks

The critical rivalry in the years ahead will not be between countries but between technical systems that countries choose to defend. Rival information infrastructure is the single most important long-term threat to international stability. The empire of the Western-inspired, but now truly global, internet isn't the only major system in which political values and information infrastructures are deeply entwined. Indeed, there are many internets. Witness the way Russians, Iranians, and Chinese use their social media in different ways. The Iranian blogosphere is full of poets.[57] The Russian blogosphere has lots of nationalists, such as the Nashi from the Seliger Forum.[58]

There will be more asymmetric conflicts, in which upstart civic leaders organize protest networks with surprising impact and the media, not just states and elites, are the targets. All political organizations, from parties and government offices to militaries and armies, will hemorrhage information. There will be more whistleblowers and defectors, and the nastier a government is, the more it will hemorrhage information about its corruption and abuses. Every dictator in the world will face embarrassing videos he cannot block and public outrage he cannot respond to. There will also be more clicktivism, half-hearted consumer activism, stillborn protests, and social movements that bring chaos to city centers but can't bring voters out on election day. Such civic engagement and nonviolent conflict will still be valuable, especially in parts of the world where ruling elites need to see that new forms of collective action are now possible. The Chinese and Russians are leading the race to build rival information infrastructures, policy environments, and cultures of technology use. What's happening in Russia and China is happening elsewhere.

But digital activism is on the rise globally, and the impact of these activist projects grows more impressive year by year. The Arab Spring involved countries where citizens used social media to create news stories that the dictators' broadcasters would have never covered. Bouazizi's self-immolation and Said's murder became the inciting incidents of uprisings because of social media. Tunisia's Ben Ali and Egypt's Mubarak were caught thoroughly off guard by social-media organization, with its heady mix of old and new activism. In Iran, the opposition Green Movement continues to deploy Facebook, and possesses eloquent bloggers to advance its cause, while the mullahs' propaganda response comes in the form of movies and broadcasts.

The surveillance leakage of technologies across jurisdictions is also something that democracies wrestle with. In democracies, the technologies we design leak to lower and lower levels of law enforcement. Such technologies can be an important part of the toolkit for fighting crime. They often start off highly regulated, with high-level access. Eventually, they trickle down to average police departments. The LAPD gets Stingray cell phone trackers that allow them to listen to calls over mobile networks.[59] Other departments are also investing in drones, even as privacy activists push back on the use of both technologies by local police.[60]

The developments can be positive in that many authoritarian regimes can't control the leakage of new technologies to politically active citizens. The devices that a regime considers consumer electronics for economic productivity get used for political conversations, repurposed and reconfigured. In authoritarian regimes, they leak out of the regime's control.

Beijing may not be able to produce the online content that the world wants to see. The Chinese government will continue to put resources into the content it wants to see. It will try to build an internet of things with access protocols that make surveillance through many devices easy. Globally, it may control more and more of the digital switches over which the world's content flows.

In an important way, China's national network is not just a subnetwork of the global internet. As a sociotechnical system, that country has become quite distinct. It was built, from the very beginning, as a tool for social control and cultural preservation. Every new device and new user extends the political reach and capacity of the network. As the internet of things evolves over most of North America and Europe, it too will be given surveillance and censorship tasks. But the greater danger to the stability of the pax technica are the privileged content and infrastructure firms that actually *want* to create subnetworks of closed content and devices.

7 BUILDING A DEMOCRACY OF OUR OWN DEVICES

The internet of things will be the most powerful political tool ever created. By 2020 there will be some thirty billion devices connected to the internet, and political power over the eight billion people on the planet will rest with the people who can control those devices. Ideally, we will all share control in responsible ways. Political clout already comes from owning or regulating mobile-phone networks, controlling the broadcast spectrum, and having the ability to shut these things down.

We have to fundamentally change the way we think of political units and order. Digital media have changed the way we use our social networks and allowed us to be political actors when we choose to be. We use technology to connect to one another, and to share stories. The state, the political party, the civic group, the citizen: these are all old categories from a predigital world. Action and reaction among governments, with occasional involvement of substate political actors, once propelled political conflict and competition. But now the interaction is continuous, and between many kinds of actors. The agency of individuals is being enhanced by the device networks of the internet of things. Increasingly, international relations will be about interpersonal relations and how devices talk to one another.

All the creative civic projects making use of device networks demonstrate that their impact depends on the power of their ideas and their effectiveness at social networking. This is in sharp contrast to nations, where power comes from the size of the population, economic wealth, or military arsenal. Civic networks are more creative, and better at deflating ideological arguments by political parties and dogmatic leaders. They can focus on problems in a sustained way, while governments and political parties have to juggle competing priorities. Competing networks exist in several forms, and to make this sociotechnical system function fairly, we need to work to strengthen the information infrastructures that have the most open standards, the widest reach, and the greatest potential for innovation.

We all need to take a more active interest in our own information security and in international affairs. We need to make sure the internet of things works for us. Program or be programmed, as hackers say. If we aren't purposeful in designing the internet of things, we'll find that those with power will make decisions using data gleaned about us, and without our informed consent. Our behavior will have more meaningful political consequences than our attitudes or aspirations.

Your Coffee Betrays You

The internet of things will be useful in many ways. Yet over the past twenty-five years we've seen many examples of how politically closed subnetworks of communities and devices degrade, while open networks thrive and evolve. Open networks share content and communicate quickly, making it easier to solve problems and take full advantage of available resources.

The internet of things will turn everyday device networks into politically valuable data. For example, pod coffeemakers have become popular sources for users' daily dose of caffeine. They have sprung up in offices, restaurants, and homes around the world. One of the most prominent distributors of these pod coffee machines is working on ways to prevent the machines from using pods made by other companies.[1] This seems like an obvious business strategy—more long-term profits can come from locking in a customer to a supply chain of coffee. Key to the process of locking in its customers to this supply chain are the chips in machines and pods that talk to each other and confirm their compatibility. If you purchase the right coffee pods, the machine will make your coffee.

It's unlikely that your coffeemaker will be put to work on some of the world's grand collective action problems, though we could imagine scenarios in which that might happen. Still, the worst way of developing the internet of things is to let it become finicky, proprietary, and materially locked down by DRM. Programming our appliances to take only the consumables approved by the manufacturers would have both economic and political consequences. Obviously it would make our economic consumption more path-dependent. The choices we have in the years ahead would be limited by the decisions we make now about our device networks. Similarly, if we construct the internet of things without asking for control over the data streams in our early device networks, we will have limited control over those streams as we add more devices. Building device networks to be closed systems, and not interoperable or hackable, encourages other manufacturers to do the same. Public leadership on an open, flexible internet of things would guide technology designers to be creative and responsible.

Digital-rights management has become one of the most legally complex and politically intractable problems of our time. Many legal experts have argued that it is a largely a problem of political culture—clear political leadership could resolve many issues.[2] So what if we allow the digital-rights management regime that was shaped by industry lobbyists to envelop the internet of things? In a world of interconnected devices, it is not impossible to imagine that your coffeemaker would shut down when you try to use coffee from your local roaster—after, of course, transmitting data about your usage patterns and your attempted hack to the other devices in your kitchen, the device's manufacturer, and the coffee industry association. Imagine a disapproving "smart" home: would consuming unapproved or incompatible products have an impact on the terms of service or generate data for government regulators?

Upending the grounds that remain in a cup of coffee is an ancient fortune-telling method. This is a magical process of divining your future and interpreting your thoughts. Increasingly, the devices attached to the internet do more than reveal our behavior. They betray our behavior and share data about us with other devices and other people we don't know. As ever more products get internet addresses, the depth of connectedness among our devices will be difficult for any one of us to keep track of.

For many years, there has been concern about network neutrality—the challenge of making sure that the owners of infrastructure on a digital network don't favor some content over others with special privileges in speed or storage. Before digital media, experts demonstrated the negative impact of media monopolies on the public sphere. Media concentration, whether in private or state hands, has always had bad effects.

In the years ahead, the ownership patterns we need to track involve more than just the producers of content and the backend superstructure of the internet. We need to be aware of how the internet of things is developing, with everything from mobile phones to digital switches. If content, search, or data-mining businesses own whole subnetworks of devices, will that create privileged content networks? Machines for the rich, machines for the poor?

The pax technica is an empire of people and devices. With sensors and activators embedded in more of the objects we produce, we must be conscious of how the constant flow of data can enrich our cultural, economic, and political lives. We must also learn from our recent experiences with politics and digital media, and be deliberate in the design of the internet of things.

This is a technical peace in the sense that the major battles may no longer be fought by militaries but by corporations with competing technical standards and a vested interest in making systems interoperable or closed. The competition over infrastructure can be fierce, and the race to fill our lives with an internet of things may involve proprietary claims, independent subnetworks of people and devices, and technologies that do not play nice with each other. The fiercest political competition in the years ahead will be over the standards-setting process for device networks. Openness does not refer just to the firmware or to increased consumer choice in electronics.

Internet Succession: Computers, Mobiles, Things

From one of the most politically peripheral corners of Mexico's Lacandon Jungle, the Zapatistas taught the world how information technology could be used to advance a social cause. More recently, democracy advocates set popular uprisings in motion across an entire region of North Africa and the Middle East. The internet involved in each of these social movements was different—the Zapatistas successfully linked their content across web-pages and computers. During the Arab Spring, activists relied on mobile phones that spread their content through text messages and social-networking applications.

Social media allow protesters to maintain nonviolent strategies. Violence by protesters gives regimes the green light to respond in kind—and authoritarian governments are usually better at violence. Social media let people build the norms of trust and reciprocity needed for successful community engagement.[3] If the internet of things consists of billions of networked devices, how will this internet be used for political protest?

During the first twenty-five years of the internet interregnum, we saw ever-increasing numbers of people using technology to solve collective action problems. The pace of problem solving increased dramatically with the evolution of social media that made mobile phones the most important devices for social connectivity. Social media have been especially good at exposing the lies of rigged elections and at bringing corruption to light. It used to be that governments produced and implemented the binding rules of social coordination and provided collective goods, ideally, in exchange. As governments—and other kinds of formal organizations—fail to provide collective goods, people use their own social networks to repair and rebuild.

Digital media use has had two important impacts on international affairs. The first is a process called connective action: social media are bridging and bonding civic networks faster than they are strengthening criminal networks. The second is a process called connective security: big data is making it easier to crowd-source information about the threats and solutions to our well-being.

If these assertions are valid, then we find ourselves with a surprising kind of stability, almost predictability, to the way political power will flow in the years ahead. This pax technica is a dynamic, systemwide stability ensured by patterns of interaction and the domains in which political power will be exercised. Internet-connected devices now mediate our political culture. If the internet of things grows as projected, it will eventually contain our political culture.

Coalitions of Western governments and technology firms control most of the world's information infrastructure and enjoy almost unchallenged influence over that infrastructure. Understanding the path of the political development of the internet gives us these premises for what the internet of things will mean for global power politics. These premises will have consequences, and we all need to appreciate that using the internet of things is a political activity. If government and industry aggressively set technology standards on their own, they make technology use political through the streams of behavioral data they will collect and act upon.

In a sense, the idea of the pax technica is straightforward. Whoever controls the largest device networks gets the most sensor data and manages the largest number of relationships between and among people and devices. The Roman Empire lasted because the Romans built roads. The British ruled the world because they built ships and robust trade alliances. The United States dominated because it had atomic bombs and engineered the post–World War II Bretton Woods system.[4] While the internet may have started off as an American technology, its current users—and developers—come from all over the world. Digital media like mobile phones and the internet are tough to monopolize and control. They are the sources and conduits of modern political power, as the infrastructure of roads, ships, and weapons have served political power in years past. But information infrastructure offers different kinds of political power to different actors, and the rules that govern modern politics have radically changed.

The standards-setting process for information technologies has become the one policy domain that, over the long term, will affect all other policy domains. Technology policy decisions affect what scientists can learn about the environment, what the World Health Organization can learn about disease vectors, what the Food and Agriculture Organization of the United Nations can learn about caloric intake, and what security officials can learn about suspicious financial transfers. Decisions about how to set up and govern information infrastructure have a path-setting impact on how scientists, public policy makers, and interested stakeholders communicate to their publics and arrive at decisions. Today's decisions about implementing the internet of things will have political resonance for generations to come.

Global technology policy is central precisely because it sets the rules by which progress in all other policy domains can be made. This means that international tensions over competing technology standards are only going to increase. Some of the most banal engineering protocols for how the internet works can have immense implications for everyday life. If the Russians, Chinese, or Iranians can put those protocols to work for their political projects, they will. Technology standards used to be left to technocrats—the experts who actually understood how the internet worked. Leaders of all stripes have seen how information technology shapes political outcomes, so they have an incentive to set the standards that serve their interests.

The World Ahead

The eighteenth-century British historian Edward Gibbon, in *The History of the Decline and Fall of the Roman Empire*, argued that "the Roman Empire was governed by absolute power, under the guidance of wisdom and virtue."[5] At this point, the networks of devices connected to the internet are allowing us to provide many kinds of governance goods for our own communities. We can't be certain that this will always be so for the internet of things. The next internet will certainly be used to express and challenge political power. Now is the time to encode the next internet with democratic virtues. More than ever, technology, including technical expertise, means political power. Political clout now comes from owning or regulating mobile-phone networks, controlling the broadcast spectrum, and having the expertise to turn off access to both.

By 2020, the majority of the global population will live under limited and fragile governments, rather than stable democratic or authoritarian ones.[6] The internet of things, more than formal governments, will be providing political structure. In 2000 only about 10 percent of the world's population was online. By 2020, most of the world's population will either be online or be economically, culturally, and politically affected by the internet of things. Three of every five of these new internet users will live in a fragile state. The things that constitute and define the internet will not be just computers and mobile phones but the objects of everyday life: lamps and refrigerators, sneakers and biosensors. Some thirty-two million smart thermostats will be installed in U.S. homes by 2020.[7] Myriad devices will consume much of the internet's bandwidth: talking among themselves, reporting on their status, and revealing the behaviors of the people in sensor range. Political bots will use much of the rest, and original, human-made content will be a fractional amount of the information that is exchanged. Data sharing among devices will be as ubiquitous, and as nearly unnoticed, as electricity.

Technology companies will become the arbiters of human rights, because it will be through their devices that we will learn of abuse, and through their devices that abuse is carried out. National security agencies around the world will be engaged in conversations about cyberdeterrence that continue to use the language of nuclear deterrence developed for an earlier era. This may not seem productive, but having political leaders fear the consequences of launching a major cyberassault is likely to result in a stable balance of power maintained and monitored through the internet of things. Instead of being caught up in an arms race, political actors will race to develop better bots. With device networks expanding rapidly, botnets will have many more nodes and passages in which to grow, and many more sensors that can be activated to surveil and manipulate public opinion.

One of the most eloquent campaigners for internet freedoms is Rebecca MacKinnon. She wrote the book on the struggles many internet users face when trying to build political futures for themselves in authoritarian countries.[8] Some people argue that information access isn't a human right, or is at most a secondary priority to more basic needs. When MacKinnon was in Tunisia for the World Summit on the Information Society in 2002—ironically, Tunisia was a dictatorship then—an audience member challenged her pleas for internet freedoms. The real priority for the world should be feeding people, clothing people, and providing housing, argued the challenger. After addressing these issues, the argument went, we could worry about information freedoms. One of MacKinnon's collaborators immediately responded: "Without freedom of speech, I can't talk about who is stealing my food."[9]

This is the reason that the internet of things must have a civic and public function. By 2020, many of us will inhabit a world of interconnected sensors that will have been embedded in everyday objects, and increasingly in our bodies. Filling our lives with such devices should not be just about making better consumer products but about giving us the ability to improve our quality of life.

The Hope and Instability of Hackers and Whistle Blowers

We've come to depend on hacktivists and whistle blowers to teach us about how this internet of things is evolving. It's easy to despise Edward Snowden and Chelsea Manning for the perceived breach of trust with their national security colleagues and the armed forces. And industry lobbyists work hard to paint activists like Aaron Swartz as miscreants.[10] But it is difficult to ignore the debates these hackers and whistle blowers set in motion. Refusing to address their questions is foolish. They risk breaching the trust of their colleagues, but earn public trust and trigger a much-needed, evidence-based public conversation about what our device networks are being used for.

Any government agency, if left unchecked and unsupervised, can violate human rights, abuse citizens, or overreach its authority in defense of its organizational interests. Throughout history, we have relied on whistle blowers to expose the bad behavior of such agencies. Valuing whistle blowers is not about encouraging rampant and unconsidered security leaks. Whistle blowers are valuable because they demonstrate how deviant government agencies have become, and the threat of public attention from a whistle blower may even deter other agencies from working beyond their public charge.

These days, whistle blowers seem to drive many of the big issues that capture headlines on technology, security, and governance. Leaks about what firms and governments can already do with devices rightly scare most of us. These days, the computer and mobile phone are the primary devices of the internet. Those are the network devices through which we work and connect to family and friends. The latest scandals typically involve the government and corporate actors gleaning information about us, usually without our informed consent, without allowing us to opt out and without providing a way to retrieve and destroy the data about us. As more of the things we manufacture are powered and networked, "inanimate" objects will be replaced by devices that talk with our other devices. They will communicate with their original manufacturer, the information services we subscribe to, national security agencies, contractors, cloud computing services, and anyone else who has broken into, or been allowed into, the data stream.

One safe inference about our future is that criminals and bots will have more devices to manipulate. The Russian Business Network has become a service that essentially provides IT support for criminal networks.[11] For a while it was openly selling a keylogging software for $150. The organization is probably behind the Storm botnet described earlier, and it actually specializes in identity theft services. The Russian government taps it for work projects. It contributes to the international market for zero-day exploits, trading in software flaws that a buyer can only use once against a device.[12]

For such dubious businesses and criminal actors, the internet of things will serve as a vast array for gathering data and a means of providing illegal information services. Coupled with the largely unregulated but not illegal markets in data about people from around the world, much of what is collected over the internet of things will be valuable—and valued—by lobbyists everywhere.[13] Denial-of-service attacks can be ordered online for between five and one hundred dollars, depending on the size of the target.[14]

Hacktivists and whistle blowers will continue to teach us the most about political actors' use of inconspicuous devices to manipulate public opinion and manage political life. The number of people in the United States who have access to sensitive information is considerable. In 2013, more than 5.1 million civilian government employees, military personnel, and contractors were eligible for some kind of security clearance.[15] Of these, around 2 million were eligible but hadn't been given access to sensitive material. The rest had been given access, and some 1.4 million citizens had very high levels of access to top-secret material.[16]

One reasonable expectation we should have of our modern democracy is that the government would not allow national security agencies to deliberately undermine internet infrastructure. Among Snowden's revelations is evidence about how the NSA has weakened the internet by deliberately introducing flaws into cryptographic systems so that its agents could read encrypted traffic.[17] The Heartbleed bug, which was publicly identified in 2014 as a software flaw exposing up to two-thirds of the world's websites, was actually identified by the NSA two years earlier. The discovery was not publicized but in fact was kept secret and actively used to gather intelligence.[18] Some officials might justify the national security advantages of allowing infrastructure flaws to persist. Healthy and stable infrastructure should not be deliberately degraded.

A broad, global, democratic agenda must now include strengthening, not sabotaging, open device networks. When officials choose specific strategic gains over long-term democratic values, they tempt potential whistle blowers to act. The security establishment is burgeoning with a new generation of data-savvy citizens who simply have different values about information openness. Such a citizen might value the chance to serve her country, have a stable job with a good salary, and work in the interesting world of intelligence. She probably has no history of civil disobedience, which would make it less likely that she would have gotten the requisite high level of security clearance. She values information openness, and is willing to risk life as she knows it by publicizing government activities online.

Most hacktivists are young, and have shown that they have a different set of values from those held by the organizations they work in—and expose. Moreover, hacktivists rarely stop their work. Srđa Popović, the Serb who in 2000 mobilized the resistance to end Slobodan Milošević's rule, went on in 2003 to train protesters for Georgia's "Rose Revolution," Ukraine's 2005 "Orange Revolution," and the Maldives' revolution in 2007, before training activists in Egypt's April 6 Movement in 2008. Popović's book *Nonviolent Struggle: 50 Crucial Points* has been downloaded thousands of times.[19]

For the presidents of countries and companies, people like Aaron Swartz, Chelsea Manning, and Julian Assange are threats to national security and the corporate bottom line. But in many networks they are heroes. Every few years, hacktivists and whistle blowers turn national security and diplomacy upside down by putting large amounts of previously secret content online.

Conservative security analysts and industry pundits often react hostilely to people who play with information technologies and exploit consumer electronics beyond designers' intent.

Still, working for a better world increasingly means forming digital clubs or putting your crypto clan to work by being creative with digital media in a political domain. Patrick Meier and his extended network of friends started a wave of crisis-mapping projects that take advantage of the energy and altruism of volunteers. Eman Abdelrahman's "We Are All Laila" project inspired a cohort of young Egyptians to think collectively about their problems and what they could do about them. Brown Moses diligently does his documentary work on Syrian munitions. Other crypto clubs—such as the Tactical Technology Collective and Citizen Lab—develop the software workarounds that let people in closed regimes get access to the internet.[20] Because these hacktivists are so well connected, governments that go after them risk upsetting extended networks of tech-savvy people. Arbitrary arrest, secret trials, and long periods of detention without charge remain a common way of going after bloggers and other online activists, even in the West.[21] Because of social media, news of arrests spreads quickly, and trials are harder to keep secret. Indeed, when a country's leaders target information infrastructure itself, leaders usually lose.

Young people in failing states take to digital media with such energy because they suffer more than their seniors. Jobs go to older adults with social connections, so youth unemployment rates are often high. Young people get restless, and they see what their friends and family overseas have. They start off by sending jokes and irreverent text messages. When pushed, they join political conversations and sign online petitions. They may not hit the streets and attack the government right away, or on their own. But when crises do emerge, political leaders increasingly find that there are surprising new information networks in place. Old, established political leaders aren't always able to remain the mandatory point of passage for information, and security services can't always keep track of how the latest device networks are being used. Social media often get used for good, especially when times are tough, because they enable people's altruism.

Firing the Social Scientists—and Training New Ones

Most social scientists have a tough time talking about what's ahead. Only a small—but growing—number are using digital media as a research tool. There's already lots of technology used in places social scientists don't look, or can't look very easily. This means that most social scientists won't be ready for the internet of things. Computer scientists have an important tool kit, but have little experience interpreting socially significant phenomena or making workable public-policy arguments. A new breed of social scientist needs to understand sociotechnical systems and analyze big data, but provide comparative context and causal narratives and demonstrate the political affordances of new device networks.

The reason we need a rapid turnover in the social sciences is that so many people have been trained to see categories that will have much less significance to the networks of devices in which we will be embedded. Moreover, the political and economic sciences have a well-evolved language for studying markets, classes, and nation-states that is quickly becoming obsolete. As I argued early in this book, most social scientists are trained to think of nation-states as representative systems—models of government defined by the relationships between people. Instead, researchers need to be trained to see the world as a system of relationships between and among people and devices.

Social scientists often begin their academic lives reading Adam Smith and Karl Marx. Smith was among the first to search for the laws of interaction in markets that communicate signals of demand and supply. Marx was the first social scientist to do systematic research—collecting large amounts of evidence in purposeful ways—into the premises and consequences of political power. Their different ways of framing social life have endured because they used a largely analogue tool kit and worked on highly aggregated categories like market, class, and nation. These categories had to be operationalized through averages and ranges. In the United States, for example, the working class might include households with an annual income between $35,000 and $75,000.[22] Or the working class might be defined by average levels of income: an average working-class man earns $57,000 and a working-class woman earns $40,000. Either approach is meant to find the most sensible general tendency among the largest number of people.

Device networks, in contrast, render detailed individual data on network tendencies. What was latent is now discoverable, and what counts as data about social life is changing. As Sandy Pentland points out, social life is made up of millions of small transactions between individuals.[23] The patterns in those individual transactions can be reduced to averages but no longer have to be. Important details about all these people and all these devices actually determine social outcomes. Traditional statisticians are shy about making causal claims with the customary tool kit. If standard maps no longer reflect real relationships on land, and existing analytical categories don't capture enough nuances, why should we stick with conventional ways of doing social science? If real-time social science data is available, and orthodox social scientists are unwilling to make strong causal claims, why would we trust their findings?

In important ways, governments no longer define citizens and manage the way they express that citizenship. Devices play that role. Modern citizenship has become a data-driven obligation. Or, more accurately, citizens increasingly make their impact through what I've called a "data shadow"—the silhouette of preferences that is cast by things like credit card records and internet use.[24] Ensuring that you get control over the data you leave behind as you move through device networks is going to become one of the most important civil liberty goals. If you defer on this responsibility to track the trackers, government and industry will set the permission levels on who gets what data about you. You don't want that, as only you really care about your privacy. Moreover, there's no guarantee that you will have access to the data about your behavior.

Putting the Civic into the Internet of Things, Domestically

In this day and age, you either set the technology standards or you follow them. Many brilliant civic projects provide governance through the open, considered, and deliberate use of the internet. So we need an internet of things that allows expression and experimentation. Brett Frischmann makes this same argument in *Infrastructure*: all public works like the internet of things should be open and nondiscriminatory.[25] We need to make sure the internet of things is designed for civic engagement. These days, it's normal for civil-society groups to have an internet strategy or a social-media strategy. Are such groups ready with a strategy for the internet of things?

Authoritarian regimes and unscrupulous politicians who stay in character will throw bots into the internet to obscure issues and muddy public opinion. Should we wait for that to happen? We need a comprehensive civic strategy for the internet of things, and we need it soon, not eventually. Why should there be a coordinated civic strategy for the development of the internet of things? What should that response be?

If the modern state is a sociotechnical system made up of people and networked devices, then we need to adapt our working definition of democracy and our expectations for what democracy can be. Democracy is a form of open society in which people in authority use the internet of things for public good and human security in ways that have been widely reviewed and publicly approved. Open societies depend on coherent collective identities, shared motivations, and opportunities to act together. Here's how I think we can build that open society.

First, the devices on the internet of things should tithe for the public good. Media historians have argued that the early newspaper industry benefited greatly from public subsidies.[26] And the countries with the best-functioning communities of investigative journalists are those that allow some public funding of civic-minded news organizations. What if the internet of things rolled out a kind of "technology tithe," in which some portion of the device's abilities were reserved for open and public use?

Ten percent of the processing time, 10 percent of the devices' information-storage capacity, and 10 percent of the bandwidth used by devices could be reserved for third-party, civic, and open applications. Public-health organizations, libraries, non-profit organizations, academics, and get-out-the vote campaigns would be allowed to make use of a certain percentage of our devices. Tech-savvy democracy advocates are already working on ways to build shadow networks that can piggyback on existing infrastructure but be immune to surveillance and censorship by profit-hungry firms and abusive governments.[27] Regardless of whether they succeed, reserving some capacity for civic life over the internet of things means creating networks that can never be blocked, filtered, or shut down. As crisis mappers have shown, disaster relief is essentially a giant logistical operation. The internet of things could be ready for duty during complex humanitarian disasters or any moment when altruism should be especially encouraged.

Second, the data produced over the internet of things should be more openly shared than that which is being produced by the current internet of mobile phones and computers. A recent study of government data sources found that most countries get lousy scores for providing data on corporate actors and non-profit groups.[28] Most governments say they promote open data. But there are actually only a few countries where you can easily find the legal name of a firm and get an official address, incorporation date, status, and other basic details.

The data collected by device networks will have to feed manufacturer and industry analysis—there's no way around that. The value chain for building the internet of things includes access to rich flows of data. If users can't opt out of being surveilled by device networks—especially by the devices they buy for themselves—users should have the ability to add to the list of organizations that can have access to data flows.

Third, people should be allowed to decide what kinds of data about their lives should qualify for the "aftermarket" in data. The United States has few rules about what firms can do with data mining. One of the rules is that a data mining firm can't profit by selling access to voter registration files. The state collects these and provides them as a public service. Political parties, civic groups, and candidates for election access the data because it helps them develop good strategy for engaging with the public. In practice, data-mining companies merge voter registration files with credit card records, public-opinion data, and other bits of information to create a rich database that can be sold. But in principle they are not supposed to make money off the voter registration files alone. Big data from devices, whether from satellites overhead or from everyday objects, tends to flow to governments and companies, and occasionally to scientists. It needs to flow to civic groups, and not simply be "open."

The dangerous approach is to allow government and industry to monopolize the use of device networks, data, and metadata. The best approach is to let many stakeholders have access to the data that people are willing to share. So the third recommendation is that people be allowed to decide what kinds of data about their political lives could also qualify for this nonprofit rule. If my voter registration data is public and useful, and others can't profit from it, I should be able to label other kinds of data about my political life that way. If I feel passionate about fair-trade coffee, and my coffee consumption data might help producers improve their products or profit margins, I should be able to prevent the maker of my coffee machine from hoarding that data.

This is a form of user control that many people might want, though many people might choose to not have any data shared with anyone. Particular citizens will probably never be able to completely opt out of the internet of things. If I'm fiscally conservative, or socially liberal, or want to do my part to slow down climate change, I should be able to share relevant data with the groups that could learn from my behavior and help me behave more responsibly. If valuable data is going to be extracted from device networks by industry and government, users should be able to share it with civic groups, hospitals, and scientists as well.

This should be expanded significantly, to cover a wide range of data that should be publicly valuable. In other words, if the government requires that firms, civic groups, and people provide data about themselves, the government should do the work of assembling that data in a basic way and providing easy access to analysts. Agencies need to think about which data should be private. Some commercial providers may figure out additional ways of packaging or analyzing data, and may charge for adding that value. But it should be illegal for political consultants, data-mining companies, and other private actors to profit from data that is shared with the state for the public good.

Fourth, each device produced for the internet of things needs to be able to report the ultimate beneficiary of the data it collects. Tax evaders, terrorists, drug cartels, and corrupt politicians don't want to keep their dirty money under their own names. So one of the most important anticorruption campaigns is against anonymous companies that are able to hide their ownership structure in layers of easily created shell companies.[29] The cruel industry behind blood diamonds, in particular, has been able to bury the identity of company owners and beneficiaries.

Unfortunately, only the most experienced data sleuths can track down their personal data and see who is using it. Given the large volumes of compromised personal records—on average each U.S. adult has had nine such records compromised—it would be impossible to fully understand who has access to data about us.[30] National-security organizations may have better digital archives of our communications than we have on our own devices. Terms-of-service agreements are increasingly complex, and there are numerous efforts to simplify them.[31] In many countries, federal agencies require food to come with simplified labels. Energy-intensive appliances have to display efficiency grades. Every device that is added to the internet of things should be tagged with up-to-date and simplified records on the beneficiaries of data flows from each device.

Occasionally, we see how much data about us exists. Spreadsheets of information surprise us with their detail, or lists of variables are published that have been compromised by a hacker, clumsy firm, or careless government office. If we cannot opt out of data collection over the internet of things, we should require that devices identify the ultimate beneficiaries of data flows. Personal identity is organized around government-issued birth certificates, passport documents, and social security numbers. Even though there's a significant industry of analysis and data miners who append our behavioral data to those records, we don't know who benefits from our data. Software protocols make it difficult to know what is collected and where it is sent. For products, the maze of engineering firms, assembly services, and intermediaries shield those beneficiaries from all but the most patient hacker. By default, device networks should have high privacy settings. The next best alternative is clearly identifiable beneficiaries of data flows.

Fifth, a number of industry groups are developing a collective conversation about how technology firms should work with governments—we need more of this. In the United States, the Global Network Initiative (GNI) has emerged as among the most important industry associations.[32] For several years, Western tech firms eager to enter China had to cooperate with that government's requests for information about democracy advocates. Between 2002 and 2004, Yahoo! supplied the Chinese government with information that helped it arrest several activists. Microsoft deleted the blog of an activist in 2005, and Google built a Chinese-based version of its search engine in 2006. Cisco began supplying the Chinese government with its internet monitoring and filtering equipment. Within a few years, pressure from Western governments and democracy activists started to have some impact. Bad press forced the industry to consider its role in the political life of tough regimes.

Some firms started behaving well, signing up for the GNI to help them coordinate their response to the world's dictators. Google stopped offering its censored search results, and both Microsoft and Google kept their email services out of easy access by China's security services. Other firms worked out complex foreign ownership arrangements that make services and technologies available in China under other names, still allowing collaboration with the Chinese government, but making the collaboration seem innocuous. So several developments could improve the likelihood of a "good" internet of things. More Western technology firms could get behind the GNI and participate in the conversations about corporate responsibility that it leads. While there are ways to ensure that healthy, public conversations about the internet of things occur within countries and trading zones like North America and Europe, there are also foreign policies that project civic values into the technology policies of other countries.

Device Networks and Foreign Affairs

Some creative thinking in foreign affairs would allow us to promote open systems through the device networks of the internet of things. Since many giant technology firms from the developing world want to be publicly traded on stock exchanges in the West, they should be held to the same ethical standards as Western firms traded on those exchanges.

Developing countries often turn to wealthy countries for loans and advice on how to implement good technology policy. All governments need to be diligent about having open public auctions for access to the public spectrum, and to continue providing extra support to women- and minority-owned businesses competing for broadcast licenses. When governments sell off access to the public spectrum through backroom deals, they usually create media titans. This happened in Mexico and India.[33] Countries that administer complex application processes, make special deals with particular firms, and allow industry lobbyists too much access to government technocrats end up with billion-dollar corruption scandals and media oligarchs. India lost between $8 billion and $20 billion from a badly managed licensing process that benefited corrupt officials and shell companies but left mobile-phone users with worse service, not better.

Governments should have open and transparent ways of allocating the public spectrum that encourage ownership diversity. Auctions are straightforward, and special credits for women-owned, minority-owned, and locally owned businesses can increase the diversity of ownership within many sectors that make use of public resources like the broadcast spectrum.

When access to the spectrum is publicly auctioned in coordination with programs to improve ownership diversity, the outcome is a transparent process with diverse stakeholders.[34] If we buy the argument that a "device tithe" could also go a long way to promoting public access to the internet of things, then perhaps 10 percent of the profits of such auctions should be set aside as a technology fund for public-interest groups. Or 10 percent of the public spectrum could be set aside for creative civic projects and device networks for disenfranchised communities.

This is not a new idea. Universal service funds have a history of constructive success. Phone company customers in urban areas of the United States had to subsidize rural service throughout much of the twentieth century.[35] Making telephone networks available to all was deemed a public good when rolling out land-line telephone infrastructure seemed like a viable engineering project. Putting the internet of things to work for the public good is also worthwhile.

Finally, wiring Africa should be a development priority. The great dangers ahead will come from competing device networks, and Chinese state enterprises are aggressively selling their wares and technology expertise at great discount to African countries in need of infrastructure. Internet access is desperately slow there, with consequences for economic, political, and cultural life.[36] In part, this is because of greedy pricing strategies from Western telecommunications firms and narrow-sighted loan conditions from international lending organizations.

Much of West Africa has limited fiber-optic links to the rest of the world. Small clusters of satellite dishes point to the sky and provide the wealthiest of customers with internet access. Satellite bandwidth supplies the Central African Republic, Chad, the Democratic Republic of the Congo, Eritrea, Guinea, Liberia, São Tomé, and the Seychelles with connections to the global internet. Many of these countries get less bandwidth than would be needed to keep a small town in the United States or Europe satisfied with connection quality. The World Bank has provided $30 million in funding for a new underwater fiber-optic cable to Sierra Leone's capital of Freetown. But going from there to the countryside, or to Africa's landlocked states, is a big step. It's not a "last mile" problem, it's a last–thousand miles problem.[37]

The solutions are out there, and the customer base is large. History has shown that it is tough for collectively managed infra-structure projects to land international backing. Africa has the most to gain with good connectivity. Communities in failed states seem to beget the most creative digital alternatives in governance. It's a battle worth fighting, and if we lose there, we all lose in many ways.

Finally, some rationality has to come to the use of export controls and information sanctions. Export controls on information technologies tend to have mixed effects. Tunisians used Google Earth to map torture centers. Yet Syrians couldn't use Google Earth until late into their civil war, because it wasn't licensed for export.[38] Once people in authoritarian regimes have widespread access to new media, it becomes tough to take the technologies away. Mubarak, having faced the digital dilemma, drove more people into the streets of Cairo when he disconnected the country's internet access. When Erdoğan, the prime minister of Turkey, launched a campaign against Twitter use and access to Google, he drove millions of people to try Twitter for the first time, set up their own Tor servers, and learn about internet censorship.

Democracies and open governments should discourage the export of some technologies and services to dictatorships, especially technologies that help a regime build its own rival internet infrastructure, conduct censorship, and surveil its citizens. Some companies, such as Canada's Netsweeper, seem to actively pursue opportunities in Somalia, Pakistan, and other failing states with bad human-rights records.[39] And other technologies and services that are blocked by export controls—such as open education platforms and MOOCs—should be allowed.

Moreover, foreign sanctions can go beyond devices to people, through travel bans. Politicians who cavort with drug lords, gambling rings, and organized crime often face travel bans. Paulo Maluf, after four decades in various political offices in Brazil, can no longer leave his country because of the money-laundering charges he would face. Venezuela's defense minister, General Rangel Silva, is on the U.S. Treasury Department's blacklist for allegedly aiding drug traffickers. We keep track of the technicians who help to train the engineers of aspiring nuclear states. Also placed under travel sanctions are people like Shigeo Nishiguchi, the top Sumiyoshi-kai leader, and his deputy, Hareaki Fukuda, two men responsible for setting policy and resolving disputes within yakuza (Japanese mafia) networks.

Perhaps it's time to track the engineers who seek to hobble the internet. At the very least, stating the obvious can have important political consequences. Naming the firms and engineers who build information infrastructures for the world's nastiest dictators might have that desirable shaming effect that has worked in the past.

What can the past twenty-five years of internet politics teach us about the internet of things we might want in the next twenty-five years? First, it makes sense to discourage the export of censorship software. Censoring pornographic websites to protect cultural values rarely works and often turns into suppression of political content. Second, we should proactively and publicly help finance the construction of information infrastructures in developing countries. Such connectivity can improve the capacity of governments to serve citizens, provide more stability and certainty to the private sector, and allow civil-society groups to work more effectively. Third, we should prepare civil-society groups for their work on and over the internet of things. Civil-society organizations often create good content over computers and mobile phones that competes with regime-vetted broadcasts, partisan propaganda, and extremist cries for attention. Public use of the internet of things will guarantee the ability of all government, corporate, party, and civic actors to check one another's behavior and benefit from connectivity in balanced ways.

How Can You Thrive in the Pax Technica?

The internet of things could become a robust civic infrastructure. A network of devices that produces open-source data that is easy for users and contributors to understand, can be locally managed, can be ethically shared, and allows people to opt out would be an information network that people would actively want to contribute to.[40] But whether or not we have a public conversation about what the internet of things could be, there are ways for us to thrive in this new empire of devices.

First, you can be a more sophisticated user. Do one thing a month to improve your tech savvy. Try out a virtual private network to see how it operates. Get a PGP key, which enables "pretty good privacy," and use it to trade one political joke with a friend.[41] Use social-networking software to map out your own Facebook networks. Check your credit history using the free services of an online credit agency. Make sure the records on you are accurate and fix anything that is not correct. Scrub your social-media profiles to make them just a little more private. One individual technology user taking these steps won't change the world. Still, you can change the way you live, protect yourself, and be more aware of how your data is being used.

Second, appreciate that devices you acquire in the years ahead have an immediate function that is useful to you and an indirect function that is useful to others. Be purposeful in choosing devices that give you control over settings. Reward the firms that treat your data responsibly and punish the ones that don't with the one thing they'll pay attention to: loss of your business. Moreover, be ready to use your internet of things for political purposes. Social media and big data have become important to modern power politics. People armed with grievances, passion, and mobile phones take down nasty dictators; the most consequential international political battles are now over the minutiae of information policy, engineering protocols, and telecommunications standards. It's tough for everyone to be engaged in technology issues all the time. We can hope that some people will be engaged on most things most of the time.

Over the past twenty-five years we've learned that almost any group that can amass informational and network resources can rise to functional prominence. The internet of things will be a powerful tool kit for governments and corporations. To ensure balance in political life, civil society groups need full access to the same tool kit. You can be a functionally prominent political actor by thoughtfully managing your internet of things.

The Promise of the Pax

The past twenty-five years of internet politics can teach us about what to expect over the next twenty-five years. The next world order will be significantly shaped by—indeed contained in—an infrastructure of devices that will be constantly talking to one another.

What will power mean in this new world order? The roles of winner and loser will go to the actors who can demonstrate truths through big data gathered over the internet of things and disseminate those truths over social media. The role of loser will go to those who cannot survive scrutiny over social media, whose lies are exposed by big data, and who do not have the skill to manipulate the internet of things.

We already have twenty-five years of experience researching the political impact of networked devices, and significant research has demonstrated both the capacities and the constraints that come with political communication mediated by digital tools. Thinking about the political impact of social media and big data so far will yield some insight into the political impact of the internet of things to come: social networks will reach even more widely and be mediated by even more devices. The data won't be big, it will be enormous.

Imagine a set of magical devices that could help inspire new entrepreneurs and link democracy advocates. These devices could help people work around a lousy postal service and bad roads that slow everyone down. In the near future, devices will come with the ability to manipulate and learn from each other. They will allow people to negotiate new business ideas and help us move money quickly and securely. You actually don't have to imagine these devices because you probably already have access to several.

It's for these reasons that some people have come to call new technologies, such as the smart mobile phone, "liberation technologies." You are about to get many more such devices, and we need to think about the world we're being liberated into. If our devices are talking to one another more, won't the data generated by them, and their role in our lives, only grow in importance?

The internet of things is a medium of increasing returns, meaning that the number of people and devices connected to it determine its value. A user has limited opportunities to do creative things with their small network of devices if no one else in the community is connected to it. As billions of new devices are connected, what will you do with the ones immediately around you?

Civic engagement has become a networked, data-intensive experience. In this book I've tried to demonstrate the great diversity of ways in which people are using digital media to change their relationships with governments and corporations. Politics used to be what happened whenever one person or organization tried to represent another person or organization. Devices will be doing much of that representative work in the years ahead, and social scientists need to stay relevant by expanding their tool kits and amending their analytical frames. From now on, politics is what happens when your devices represent you in the pax technica.

NOTES

A digital object identifier (DOI) gives an electronic document a unique number. Whenever a publisher has assigned one of these sources a DOI, a number has been provided. If you search a library database or the internet for a source by using this DOI number, you will quickly find the correct document.

Preface

1. Twenty-six billion devices: "Gartner Says the Internet of Things Installed Base Will Grow to 26 Billion Units by 2020," *Gartner.com*, December 12, 2013, accessed September 30, 2014, http://www.gartner.com/newsroom/id/2636073; thirty billion connected devices: ABI Research, "More Than 30 Billion Devices Will Wirelessly Connect to the Internet of Everything in 2020," May 9, 2013, accessed September 30, 2014, https://www.abiresearch.com/press/more-than-30-billion-devices-will-wirelessly-conne; fifty billion devices and objects: "The Internet of Things," *Cisco*, accessed June 16, 2014, http://share.cisco.com/internet-of-things.html; "nanosats": "Space: The Next Startup Frontier," *Economist*, June 7, 2014, accessed September 30, 2014, http://www.economist.com/news/leaders/21603441-where-nanosats-boldly-go-new-businesses-will-followunless-they-are-smothered-excessive; OECD, *Building Blocks for Smart Networks*, OECD Digital Economy Papers (Paris: Organisation for Economic Co-operation and Development, January 17, 2013), accessed September 30, 2014, http://www.oecd-ilibrary.org/content/workingpaper/5k4dkhvnzv35-en.
2. "The Internet of Things."
3. Marcus Wohlsen, "What Google Really Gets Out of Buying Nest for $3.2 Billion," *Wired*, January 14, 2014, accessed September 30, 2014,

http://www.wired.com/2014/01/googles-3-billion-nest-buy-finally
-make-internet-things-real-us/.

1. Empire of Connected Things

1. *Internet Census 2012: Port Scanning /0 Using Insecure Embedded Devices*, 2012, accessed September 15, 2014, http://internetcensus2012.bitbucket.org/paper.html.

2. Edith Penrose and Christos Pitelis, *The Theory of the Growth of the Firm* (Oxford: Oxford University Press, 2009); "Edith Penrose," *Wikipedia*, accessed June 23, 2014, http://en.wikipedia.org/wiki/Edith_Penrose.

3. "Hudson's Bay Company," *Wikipedia*, accessed June 15, 2014, http://en.wikipedia.org/wiki/Hudson's_Bay_Company; "East India Company (English Trading Company)," *Encyclopedia Britannica*, accessed June 16, 2014, http://www.britannica.com/EBchecked/topic/176643/East-India-Company.

4. "ITU: Committed to Connecting the World," accessed June 16, 2014, http://www.itu.int/.

5. "Internet Users in the World," *Internet World Stats: Usage and Population Statistics*, June 30, 2012, http://www.internetworldstats.com/stats.htm.

6. "Hooking up," *Economist*, January 31, 2013, accessed September 30, 2014, http://www.economist.com/news/international/21571126-new-data-flows-highlight-relative-decline-west-hooking-up.

7. Anne-Marie Slaughter, *A New World Order* (Princeton: Princeton University Press, 2009).

8. World Affairs Council, "Press Conference" (Regent Beverly Wilshire Hotel, April 19, 1994); Francis Fukuyama, *The End of History and the Last Man* (New York: Simon and Schuster, 2006); G. John Ikenberry, "The Myth of Post–Cold War Chaos," *Foreign Affairs* 75, no. 3 (May 1996): 79–91.

9. James Ball, "Meet the Seven People Who Hold the Keys to Worldwide Internet Security," *Guardian*, February 28, 2014, accessed September 30, 2014, http://www.theguardian.com/technology/2014/feb/28/seven-people-keys-worldwide-internet-security-web; "Internet Society," accessed June 16, 2014, http://www.internetsociety.org/; "ICANN," accessed June 16, 2014, https://www.icann.org/.

10. Harold Adams Innis, *Empire and Communications* (Lanham, MD: Rowman and Littlefield, 2007); Marshall McLuhan, *The Gutenberg Galaxy: The Making of Typographic Man* (Toronto: University of Toronto Press, 2011).

11. "Twitter Saves Lives in Mexico," *Americas Quarterly*, accessed September 10, 2014, http://www.americasquarterly.org/node/2576.

12. "Bloqueos," *Google Maps*, accessed June 16, 2014, https://maps.google.com/maps/ms?ie=UTF8&hl=en&msa=0&msid=117897461681645229195.0004822dd7b250a595b99&ll=25.705578,-100.26947&spn=0.224894,0.308647&z=12&dg=feature.

13. "Superstorm Sandy: NYC," *Google Maps*, accessed June 16, 2014, http://google.org/crisismap/2012-sandy-nyc?hl=en&llbox=40.8579,40.5237,-73.9334,-74.3728&t=roadmap&layers=layer1,layer0,8,9,1330918331511,5&promoted.

14. "How to Use Technology to Counter Rumors During Crises: Anecdotes from Kyrgyzstan," *iRevolution*, March 26, 2011, accessed June 30, 2014, http://irevolution.net/2011/03/26/technology-to-counter-rumors/. Short Message Service (SMS) is a tool that allows for text messages between telephones.

15. "Rassd News Network (RNN)," *Wikipedia*, accessed May 21, 2014, http://en.wikipedia.org/wiki/Rassd_News_Network_(RNN).

16. "WITNESS: Cameras Everywhere Report," 2011, accessed September 30, 2014, http://www3.witness.org/cameras-everywhere/report-2011.

17. Andrés Monroy-Hernández et al., "The New War Correspondents: The Rise of Civic Media Curation in Urban Warfare," in *Proceedings of the 2013 Conference on Computer Supported Cooperative Work* (ACM, 2013), 1443–52, doi:10.1145/2441776.2441938, http://dl.acm.org/citation.cfm?id=2441938.

18. Joe Davidson, "Too Many People with Security Clearances, but Cuts Could Help Some Feds, Hurt Others," *Washington Post*, March 20, 2014, accessed September 30, 2014, http://www.washingtonpost.com/politics/federal_government/too-many-people-with-security-clearances-but-cuts-could-help-some-feds-hurt-others/2014/03/20/1f1d011a-b05e-11e3-a49e-76adc9210f19_story.html.

19. Dana Priest and William M. Arkin, "A Hidden World, Growing Beyond Control," *Washington Post*, 2010, accessed September 30, 2014, http://projects.washingtonpost.com/top-secret-america/articles/a-hidden-world-growing-beyond-control/.

20. Michael Eisen, "What Exactly Are the NSA's 'Groundbreaking Crypt-analytic Capabilities'?" *WIRED*, September 4, 2013, accessed September 30, 2014, http://www.wired.com/2013/09/black-budget-what-exactly-are-the-nsas-cryptanalytic-capabilities/.

21. Declan McCullagh, "How the U.S. Forces Net Firms to Cooperate on Surveillance," CNET, July 12, 2013, accessed September 30, 2014, http://www.cnet.com/news/how-the-u-s-forces-net-firms-to-cooperate-on-surveillance/.

22. Kim Zetter, "Google Takes on Rare Fight Against National Security Letters," *Wired*, April 4, 2013, accessed September 30, 2014, http://www.wired.com/2013/04/google-fights-nsl/.

23. "Lavabit," accessed June 16, 2014, http://lavabit.com/; "Silent Circle," accessed June 16, 2014, http://silentcircle.com/.

24. Author's calculations based on the transparency reports available from Facebook (https://govtrequests.facebook.com/, accessed September 24, 2014), Google (http://www.google.com/transparencyreport/, accessed September 24, 2014), and Twitter (https://transparency.twitter.com, accessed September 21, 2014).

25. Anas Qtiesh, "Spam Bots Flooding Twitter to Drown Info About #Syria Protests," *Global Voices Advocacy*, April 18, 2011, accessed September 30, 2014, http://advocacy.globalvoicesonline.org/2011/04/18/spam-bots-flooding-twitter-to-drown-info-about-syria-protests/; Neal Ungerleider, "Behind the Mystery of Spam Tweets Clogging Syrian Protesters' Streams," *Fast Company*, April 21, 2011, accessed September 30, 2014, http://www.fastcompany.com/1748827/behind-mystery-spam-tweets-clogging-syrian-protesters-streams.

26. "Who Is Using EGHNA Media Server," *EGHNA Media Server*, accessed September 30, 2014, http://media.eghna.com/success_stories.

27. "A Call to Harm: New Malware Attacks Target the Syrian Opposition," *Citizen Lab*, June 21, 2013, accessed September 30, 2014, https://citizenlab.org/2013/06/a-call-to-harm/.

28. Alex Cheng and Mark Evans, *Inside Twitter: An In-Depth Look at the 5% of Most Active Users* (Toronto: Sysomos, August 2009), accessed September 30, 2014, http://www.sysomos.com/insidetwitter/mostactiveusers/.

29. Brian Krebs, "Twitter Bots Target Tibetan Protests," *Krebs on Security*, March 20, 2012, accessed September 30, 2014, http://krebsonsecurity.com/2012/03/twitter-bots-target-tibetan-protests/.

30. Mike Orcutt, "Twitter Mischief Plagues Mexico's Election," *MIT Technology Review*, June 21, 2012, accessed September 30, 2014, http://www.technologyreview.com/news/428286/twitter-mischief-plagues-mexicos-election/.

31. Inside Croydon, "Jasper Admits to Using Twitter Bots to Drive Election Bid," *Inside Croydon*, accessed September 30, 2014, http://insidecroydon.com/2012/11/26/jasper-admits-to-using-twitter-bots-to-drive-election-bid/.

32. Jillian C. York, "Dangerous Social Media Games," *Al Jazeera*, January 13, 2012, accessed September 30, 2014, http://www.aljazeera.com/indepth/opinion/2012/01/20121111642310699.html.

33. "Joint Threat Research Intelligence Group," *Wikipedia*, accessed June 20, 2014, http://en.m.wikipedia.org/wiki/Joint_Threat_Research_Intelligence_Group.

34. Associated Press, "US Secretly Created 'Cuban Twitter' to Stir Unrest and Undermine Government," *Guardian*, April 3, 2014, accessed September 30, 2014, http://www.theguardian.com/world/2014/apr/03/us-cuban-twitter-zunzuneo-stir-unrest.

35. "Storm Botnet," *Wikipedia*, accessed June 30, 2014, http://en.wikipedia.org/wiki/Storm_botnet.

36. "Kraken Botnet," *Wikipedia*, accessed June 19, 2014, http://en.wikipedia.org/wiki/Kraken_botnet.

37. "The Spamhaus Project," accessed June 20, 2014, http://www.spamhaus.org/.

38. Raphael Satter, "Spamhaus Hit with 'Largest Publicly Announced DDoS Attack' Ever, Affecting Internet Users Worldwide," *Huffington Post*, March 27, 2013, accessed September 30, 2014, http://www.huffingtonpost.com/2013/03/27/spamhaus-cyber-attack_n_2963632.html?utm_hp_ref=tw.

39. "ITU"; "ICANN"; "IGF," *Internet Governance Forum*, accessed June 20, 2014, http://www.intgovforum.org/cms/.

40. Philip N. Howard, *New Media Campaigns and the Managed Citizen* (New York: Cambridge University Press, 2005).

41. David Karpf, *The MoveOn Effect: The Unexpected Transformation of American Political Advocacy* (New York: Oxford University Press, 2012); Daniel Kreiss, *Taking Our Country Back: The Crafting of Networked Politics from Howard Dean to Barack Obama* (New York: Oxford University Press, 2012).

42. Mohamed El Dahshan, "Concerned Citizens and Bounty Hunters: The Lebanese Army Has an App for You," *Foreign Policy Blogs*, September 6, 2013, accessed September 30, 2014, http://transitions.foreignpolicy .com/posts/2013/09/06/concerned_citizens_and_bounty_hunters_ the_lebanese_army_has_an_app_for_you.

43. Tim Fernholz, "The Secret Financial Market Only Robots Can See," *Quartz*, September 17, 2013, accessed September 30, 2014, http://qz.com/ 124721/the-secret-financial-market-only-robots-can-see/.

2. Internet Interregnum

1. Philip N. Howard and Muzammil M. Hussain, *Democracy's Fourth Wave? Digital Media and the Arab Spring* (New York: Oxford University Press, 2013).

2. Mandiant, *APT1: Exposing One of China's Cyber Espionage Units* (Alexandria, VA: Mandiant, March 2013), accessed September 30, 2014, http:// intelreport.mandiant.com/Mandiant_APT1_Report.pdf.

3. Andy Greenberg, "Researchers Name Three Hackers Tied to One Of China's Most Active Military Intrusion Teams," *Forbes*, February 19, 2013, accessed September 30, 2014, http://www.forbes.com/sites/ andygreenberg/2013/02/19/researchers-name-three-hackers-tied-to -one-of-chinas-most-active-military-intrusion-teams/.

4. "Getting Ugly: If China Wants Respect Abroad, It Must Rein In Its Hackers," *Economist*, February 21, 2013, accessed September 30, 2014, http://www.economist.com/news/leaders/21572200-if-china-wants -respect-abroad-it-must-rein-its-hackers-getting-ugly.

5. Lisa Ferguson et al., "Netizen Report: Tibetan Internet Users Targeted with Malware," *Global Voices Advocacy*, April 9, 2013, accessed Septem-

ber 30, 2014, http://advocacy.globalvoicesonline.org/2013/04/09/netizen-report-tibetan-internet-users-targeted-with-malware/.

6. Chris Strom, Dave Michaels, and Eric Engleman, "Cyberattacks Abound Yet Companies Tell SEC Losses Are Few," *Bloomberg*, April 3, 2013, accessed September 30, 2014, http://www.bloomberg.com/news/2013-04-04/cyberattacks-abound-yet-companies-tell-sec-losses-are-few.html.

7. "Smoking Gun," *Economist*, February 21, 2013, accessed September 30, 2014, http://www.economist.com/news/china/21572228-evidence-mounting-chinas-government-sponsoring-cybertheft-western-corporate.

8. Erik Kirschbaum, "Snowden Says NSA Engages in Industrial Espionage: TV," *Reuters*, January 26, 2014, accessed September 30, 2014, http://www.reuters.com/article/2014/01/26/us-security-snowden-germany-idUSBREA0P0DE20140126.

9. Nicole Perlroth, "Cyberattack on Saudi Oil Firm Disquiets U.S.," *The New York Times*, October 24, 2012, accessed September 30, 2014, http://www.nytimes.com/2012/10/24/business/global/cyberattack-on-saudi-oil-firm-disquiets-us.html.

10. "Stuxnet," *Wikipedia*, accessed June 30, 2014, http://en.wikipedia.org/wiki/Stuxnet.

11. Nicole Perlroth, "Virus Seeking Bank Data Is Tied to Attack on Iran," *Bits*, August 9, 2012, http://bits.blogs.nytimes.com/2012/08/09/researchers-find-possible-state-sponsored-virus-in-mideast/.

12. Federal Bureau of Investigation, "Wanted by the FBI: Wang Dong," *FBI*, accessed June 30, 2014, http://www.fbi.gov/wanted/cyber/wang-dong.

13. William J. Dobson, *The Dictator's Learning Curve: Inside the Global Battle for Democracy* (New York: Random House, 2012).

14. "Obama Upbraids China on Hack Attacks," *BBC News*, March 13, 2013, accessed September 30, 2014, http://www.bbc.co.uk/news/world-us-canada-21772596.

15. Christian Sandvig, "Corrupt Personalization," *Social Media Collective*, accessed July 5, 2014, http://socialmediacollective.org/2014/06/26/corrupt-personalization/; Jonathan Zittrain, "Facebook Could Decide

an Election Without Anyone Ever Finding Out," *New Republic*, June 1, 2014, accessed September 30, 2014, http://www.newrepublic.com/article/117878/information-fiduciary-solution-facebook-digital-gerrymandering.

16. "Anonymous (group)," *Wikipedia*, accessed June 30, 2014, http://en.wikipedia.org/wiki/Anonymous_(group).

17. Nicole Perlroth, "Hackers Take Down World Cup Site in Brazil," *Bits*, June 20, 2014, accessed September 30, 2014, http://bits.blogs.nytimes.com/2014/06/20/hackers-take-down-world-cup-site-in-brazil/.

18. "Hack into Child Porn Sites Instead, DOJ Urges Hacktivists," *GMA News Online*, October 10, 2012, accessed September 30, 2014, http://www.gmanetwork.com/news/story/277629/scitech/technology/hack-into-child-porn-sites-instead-doj-urges-hacktivists.

19. Malcolm Gladwell, "Small Change," *New Yorker*, October 4, 2010, http://www.newyorker.com/magazine/2010/10/04/small-change-3; Evgeny Morozov, "The Folly of Kindle Diplomacy," *Slate*, June 21, 2012, http://www.slate.com/articles/technology/future_tense/2012/06/state_department_s_amazon_kindle_plan_won_t_help_dissidents_.html; Gina Neff and Peter Nagy, "Imagined Affordances: Reconstructing a Keyword for Technology Studies," Center for Media, Data, and Society, Central European University, Working Paper 2014.2, September 2014, accessed September 30, 2014.

20. Aaron Smith, *Civic Engagement in a Digital Age* (Washington, DC: Pew Research, April 2013), accessed September 12, 2014, http://www.pewinternet.org/2013/04/25/civic-engagement-in-the-digital-age/.

21. Tim Bradshaw, "YouTube Reaches Billion Users Milestone," *Financial Times*, March 21, 2013, accessed September 30, 2014, http://www.ft.com/intl/cms/s/0/8f06331a-91ca-11e2-b4c9-00144feabdc0.html.

22. "Syria Internet Usage and Telecommunications Report," *Internet World Stats*, accessed June 20, 2014, http://www.internetworldstats.com/me/sy.htm.

23. Open Signal, Global Network Maps, accessed September 24, 2014, http://opensignal.com/networks/.

24. Cisco, *Cisco Visual Networking Index: Global Mobile Data Traffic Forecast Update, 2013–2018* (San Jose, CA: Cisco, February 2014), accessed Septem-

ber 30, 2014, http://cisco.com/c/en/us/solutions/collateral/service
-provider/visual-networking-index-vni/white_paper_c11-520862
.html.

25. Larry Diamond, "Why Are There No Arab Democracies?" *Journal of Democracy* 21, no. 1 (2010): 93–112.

26. Howard and Hussain, *Democracy's Fourth Wave?*

27. Francis Fukuyama, *The End of History and the Last Man*, reissue ed. (New York: Free Press, 2006).

28. Clive Southey, "The Staples Thesis, Common Property, and Homesteading," *Canadian Journal of Economics* 11, no. 3 (1978): 547–59, doi:10 .2307/134323.

29. Lita Person, *Mobile Wallet (NFC, Digital Wallet) Market (Applications, Mode of Payment, Stakeholders, and Geography)—Global Share, Size, Industry Analysis, Trends, Opportunities, Growth, and Forecast, 2012–2020* (Portland, OR: Allied Market Research, November 2013), accessed September 30, 2014, http://www.alliedmarketresearch.com/mobile-wallet-market; Marion Williams, "The Regulatory Tension over Mobile Money," *Australian Banking and Finance*, February 17, 2014, accessed September 30, 2014, http://www.australianbankingfinance.com/banking/the-regulatory -tension-over-mobile-money/.

30. "University of Cumbria Becomes First in World to Accept Tuition Fees in Bitcoin," *India Today*, January 22, 2014, accessed September 30, 2014, http://indiatoday.intoday.in/story/british-university-to-accept -tuition-fees-in-bitcoin/1/339087.html.

31. Philip N. Howard and Nimah Mazaheri, "Telecommunications Reform, Internet Use, and Mobile Phone Adoption in the Developing World," *World Development* 37, no. 7 (2009): 1159–69, doi:10.1016/ j.worlddev.2008.12.005.

32. "The Big Mobile-Phone Reset," *Economist*, September 7, 2013, accessed September 30, 2014, http://www.economist.com/news/ business/21585006-weeks-two-telecoms-deals-will-be-followed -others-industry-undergoes-big.

33. @IDFSpokesperson, microblog, *Twitter*, November 14, 2012, accessed September 30, 2014, https://twitter.com/IDFSpokesperson/status/ 268722403989925888.

34. @IDFSpokesperson, microblog, *Twitter*, November 14, 2012, accessed September 30, 2014, https://twitter.com/IDFSpokesperson/status/268780918209118208.

35. John Timpane, "Israel vs. Hamas: The First Social-Media War," *Philly.com*, October 17, 2012, accessed September 30, 2014, http://articles.philly.com/2012-11-17/news/35157659_1_hamas-tweets-alqassambrigade.

36. Ian Steadman, "Big Data, Language and the Death of the Theorist," *Wired UK*, January 25, 2013, accessed September 30, 2014, http://www.wired.co.uk/news/archive/2013-01/25/big-data-end-of-theory.

37. David Axe, "With Drones and Satellites, U.S. Zeroed in on Bin Laden," *Wired*, May 3, 2011, accessed September 30, 2014, http://www.wired.com/2011/05/with-drones-and-satellites-u-s-zeroed-in-on-bin-laden/.

38. Arthur L. Stinchcombe, "Ending Revolutions and Building New Governments," *Annual Review of Political Science* 2, no. 1 (1999): 49–73, doi:10.1146/annurev.polisci.2.1.49.

39. Philip N. Howard, *The Digital Origins of Dictatorship and Democracy: Information Technology and Political Islam* (New York: Oxford University Press, 2010).

40. Yoani Sánchez, *Generación Y*, accessed June 20, 2014, http://generacionyen.wordpress.com/.

41. Eric Schmidt and Jared Cohen, "The Digital Disruption: Connectivity and the Diffusion of Power," *Foreign Affairs* 89 (November 2010): 79.

42. Katherine Maher, "The New Westphalian Web," *Foreign Policy*, February 25, 2013, accessed September 30, 2014, http://www.foreignpolicy.com/articles/2013/02/25/the_new_westphalian_web.

43. Bruce Schneier, "Feudal Security," *Schneier on Security*, December 3, 2012, accessed September 30, 2014, https://www.schneier.com/blog/archives/2012/12/feudal_sec.html.

3. New Maps for the New World

1. "United Nations Stabilisation Mission in Haiti," *Wikipedia*, accessed June 18, 2014, http://en.wikipedia.org/wiki/United_Nations_Stabilisation_Mission_in_Haiti.

2. "The Rise of the Humanitarian Drone: Giving Content to an Emerging Concept," *iRevolution*, June 30, 2014, accessed September 30, 2014, http://irevolution.net/2014/06/30/rise-of-humanitarian-uav/.

3. Oran R. Young, "Political Leadership and Regime Formation: On the Development of Institutions in International Society," *International Organization* 45, no. 03 (1991): 281–308, doi: 10.1017/S0020818300033117.

4. Latifa Al-Zayyat, *The Open Door* (Cairo: American University in Cairo Press, 2002).

5. Mohamed Hossam Ismail, "Laila's Soft Screaming: A Discourse Analysis of Cyber-Feminist Resistance on the Egyptian Women Blogsphere," *Kolena Laila*, June 9, 2010, accessed September 30, 2014, http://kolenalaila.com/en/.

6. Doug McAdam, *Political Process and the Development of Black Insurgency, 1930–1970* (Chicago: University of Chicago Press, 1982).

7. "Bring Back Our Girls," *Facebook*, accessed June 30, 2014, https://www.facebook.com/bringbackourgirls; "Kony 2012," *Know Your Meme*, accessed June 30, 2014, http://knowyourmeme.com/memes/events/kony-2012.

8. Lee Rainie et al., "The Viral Kony 2012 Video," Pew Research Internet Project (Washington, DC: Pew Research, March 15, 2012), accessed October 10, 2014, http://pewinternet.org/Reports/2012/Kony-2012-Video/Main-report.aspx.

9. Jarrett Murphy, "1 Billion Live in Slums," *CBS News*, October 8, 2003, accessed September 30, 2014, http://www.cbsnews.com/news/1-billion-live-in-slums/.

10. "Global Issues: Refugees," *UN Global Issues*, accessed June 20, 2014, http://www.un.org/en/globalissues/refugees/; Imogen Foulkes, "Global Refugee Figures Highest Since WW2, UN Says," *News*, June 20, 2014, accessed September 30, 2014, http://www.bbc.com/news/world-27921938.

11. Howard Rheingold, *Smart Mobs: The Next Social Revolution* (New York: Basic, 2003).

12. F. Edwards, Philip N. Howard, and Mary Joyce, *Digital Activism and Non-Violent Conflict* (Seattle: Digital Activism Research Project, November

2013), accessed September 30, 2014, http://digital-activism.org/2013/11/report-on-digital-activism-and-non-violent-conflict/.

13. Jason Motlagh, "Protesters Broaden Tactics as Belarus Cracks Down," *Christian Science Monitor*, July 12, 2011, http://www.csmonitor.com/World/Europe/2011/0712/Protesters-broaden-tactics-as-Belarus-cracks-down.

14. Oksana Grytsenko, "Ukrainians Crowdfund to Raise Cash for 'People's Drone' to Help Outgunned Army," *Guardian*, June 29, 2014, accessed September 30, 2014, http://www.theguardian.com/world/2014/jun/29/outgunned-ukrainian-army-crowdfunding-people-drone.

15. "Map Kibera," accessed June 20, 2014, http://mapkibera.org/.

16. Brian Ekdale, "Slum Tourism in Kibera: Education or Exploitation?" *Brian Ekdale's Blog*, July 13, 2010, accessed September 30, 2014, http://www.brianekdale.com/slum-tourism-in-kibera-education-or-exploitation/.

17. Robert Neuwirth, *The Hidden World of Shadow Cities*, TEDGlobal, 2005, accessed September 30, 2014, http://www.ted.com/talks/robert_neuwirth_on_our_shadow_cities.

18. "Spatial Collective," accessed June 20, 2014, http://spatialcollective.com/.

19. Philip N. Howard, "The Dictator's Dead Pool for 2013: Will China's Investments Pay Off in Political Clout?" *Huffington Post*, December 31, 2012, accessed September 30, 2014, http://www.huffingtonpost.com/philip-n-howard/the-dictators-dead-pool-for-2013-will-chinas-investments-pay-off-in-political-clout_b_2374668.html.

20. Carter Center, *Study Mission to the October 7, 2012 Presidential Election in Venezuela* (Atlanta: Carter Center, October 2012), accessed September 30, 2014, http://www.cartercenter.org/resources/pdfs/news/peace_publications/election_reports/venezuela-2012-election-study-mission-final-rpt.pdf.

21. John P. Sullivan, *From Drug Wars to Criminal Insurgency: Mexican Cartels, Criminal Enclaves, and Criminal Insurgency in Mexico and Central America, and Their Implications for Global Security* (Paris: Fondation Maison des sciences de l'homme, January 2012).

22. Ibid.

23. Vincenzo Bove, "Opium Market, Revenue Opportunities, and Insurgency in Afghanistan's Provinces," Essex: University of Essex, 2011, 1–27.

24. Benjamin A. Olken and Patrick Barron, *The Simple Economics of Extortion: Evidence from Trucking in Aceh* (Washington, DC: National Bureau of Economic Research, June 2007), accessed September 30, 2014, http://www.nber.org/papers/w13145.

25. Moises Naim, "Mafia States: Organized Crime Takes Office," *Foreign Affairs* 91 (May 2012): 112.

26. Ibid.

27. Symantec, *Norton Report 2013: Cost per Cybercrime Victim Up 50 Percent* (Mountain View, CA: Symantec, October 2013), accessed September 30, 2014, http://www.symantec.com/about/news/release/article.jsp?prid=20131001_01.

28. Luke Harding, "WikiLeaks Cables: Russian Government 'Using Mafia for Its Dirty Work,'" *Guardian*, December 1, 2010, accessed September 30, 2014, http://www.theguardian.com/world/2010/dec/01/wikileaks-cable-spain-russian-mafia.

29. Homi Kharas and Andrew Rogerson, *Horizon 2025: Creative Destruction in the Aid Industry* (London: Overseas Development Institute, July 2012), accessed September 30, 2014, http://www.aidmonitor.org.np/reports/horizon%202012.pdf.

30. "Where Life Is Cheap and Talk Is Loose," *Economist*, March 17, 2011, accessed September 30, 2014, http://www.economist.com/node/18396240.

31. "Hung, Drawn, and Quartered," *Economist*, November 10, 2012, accessed September 30, 2014, http://www.economist.com/news/international/21565927-better-deterrents-are-putting-somali-pirates-business-under-strain-hung-drawn-and.

32. National Democratic Institute, "Election Stories Unfold on a Map," *National Democratic Institute*, Spring 2011, accessed September 30, 2014, https://www.ndi.org/election-stories-unfold-on-a-map.

33. *Voter Fraud Clip*, 2012, accessed September 30, 2014, http://youtu.be/jxf-nRTDvGQ.

34. "Crowdmap," *Ushahidi*, accessed June 30, 2014, http://www.ushahidi.com/product/crowdmap/.

35. "Ushahidi Haiti Project Map," *National Geographic*, accessed June 27, 2014, http://newswatch.nationalgeographic.com/2012/07/02/crisis

-mapping-haiti/uhp2/; "Russian Fires," accessed June 27, 2014, http://russian-fires.ru/; "LizaAlert," accessed June 27, 2014, http://lizaalert.org/.

36. Joel S. Migdal, *Strong Societies and Weak States: State-Society Relations and State Capabilities in the Third World* (Princeton: Princeton University Press, 1988).

37. World Bank, "Mobile Payments Go Viral: M-Pesa in Kenya," *Yes Africa Can: Stories from a Dynamic Continent* (Washington, DC: World Bank, January 2013), accessed September 30, 2014, http://web.worldbank.org/WBSITE/EXTERNAL/COUNTRIES/AFRICAEXT/0,,contentMDK:2255 1641~pagePK:146736~piPK:146830~theSitePK:258644,00.html.

38. Philip N. Howard, "If Your Government Fails, Can You Create a New One with Your Phone?" *The Atlantic*, July 31, 2013, accessed September 30, 2014, http://www.theatlantic.com/international/archive/2013/07/if-your-government-fails-can-you-create-a-new-one-with-your-phone/278216/.

39. Ibid.

40. Ibid.

41. Sara Corbett, "Can the Cellphone Help End Global Poverty?" *New York Times*, April 13, 2008, accessed September 30, 2014, http://www.nytimes.com/2008/04/13/magazine/13anthropology-t.html.

42. Human Rights Watch, *You Can Die Any Time: Death Squad Killings in Mindanao* (New York: Human Rights Watch, April 2009), accessed September 30, 2014, http://www.hrw.org/sites/default/files/reports/philippines0409webwcover_0.pdf; "Davao Death Squads," Wikipedia, accessed September 14, 2014, http://en.wikipedia.org/wiki/Davao_death_squads.

43. "The Project on Extrajudicial Executions," accessed June 27, 2014, http://www.extrajudicialexecutions.org/.

44. Peter van der Windt, "From Crowdsourcing to Crowdseeding: The Cutting Edge of Empowerment?" in *Bits and Atoms: Information and Communication Technology in Areas of Limited Statehood*, ed. Steven Livingston and Gregor Walter-Drop (New York: Oxford University Press, 2013), 144—56.

45. "Communities @ Risk: Targeted Digital Threats Against Civil Society," Citizen Lab, November 11, 2014, accessed January 4, 2015, https://citizenlab.org/2014/11/civil-society-organizations-face-onslaught-persistent-computer-espionage-attacks/.

46. Computational Propaganda, accessed January 4, 2015, http://political bots.org/.

4. Five Premises for the Pax Technica

1. Alexis de Tocqueville, *The Recollections of Alexis de Tocqueville*, trans. Alexander Teixeira de Mattos (n.p.: Project Gutenberg, 2011), accessed September 30, 2014, http://www.gutenberg.org/ebooks/37892.

2. Steven Pfaff, *Exit-Voice Dynamics and the Collapse of East Germany: The Crisis of Leninism and the Revolution of 1989* (Durham, NC: Duke University Press, 2006).

3. Nancy Dahlberg, "It's Prime Time for Hackathons and Other Technology Events," *Miami Herald*, February 10, 2014, accessed September 30, 2014, http://miamiherald.typepad.com/the-starting-gate/2014/02/miguel-chateloin-participates-in-the-hackathon-for-cuba-event-on-feb-1-at-the-lab-miami-hackathons-are-beco.html; Will Mari, "Hacking for Jesus: Top Projects from 'Code for the Kingdom,' a Faith-Based Hackathon for Spiritual Geeks," *GeekWire*, March 24, 2014, accessed September 30, 2014, http://www.geekwire.com/2014/faith-based-hackathon-meets-impact-hub/.

4. Jon Kolko, *Wicked Problems: Problems Worth Solving* (Austin: Ac4d, 2012).

5. Bruce Schneier, "The NSA Is Commandeering the Internet," *The Atlantic*, August 12, 2013, accessed September 30, 2014, http://www.theatlantic.com/technology/archive/2013/08/the-nsa-is-commandeering-the-internet/278572/.

6. "Yahya Ayyash," *Wikipedia*, accessed June 25, 2014, http://en.wikipedia.org/wiki/yahya_Ayyash.

7. Kat Hannaford, "How Osama Bin Laden Was Found," *Gizmodo*, May 2, 2011, accessed September 30, 2014, http://gizmodo.com/5797588/how-osama-bin-laden-was-found.

8. "Euromaidan," *Wikipedia,* accessed June 30, 2014, http://en.wikipedia .org/wiki/Euromaidan.

9. Heather Murphy, "Ominous Text Message Sent to Protesters in Kiev Sends Chills Around the Internet," *Lede,* January 22, 2014, accessed September 30, 2014, http://thelede.blogs.nytimes.com/2014/01/22/ ominous-text-message-sent-to-protesters-in-kiev-sends-chills-around -the-internet/.

10. Jonathan Fildes, "MEPs Condemn Iran 'Surveillance,'" BBC, February 11, 2010, accessed September 30, 2014, http://news.bbc.co.uk/2/ hi/8511035.stm.

11. Andrei Aliaksandrau and Alaksiej Lavoncyk, "Belarus: Pulling the Plug," *Xindex: The Voice of Free Expression* (Budapest, HU, January 2013), accessed September 30, 2014, http://www.indexoncensorship.org/wp -content/uploads/2013/01/IDX_Belarus_ENG_WebRes.pdf.

12. "Stuxnet," *Wikipedia,* accessed June 30, 2014, http://en.wikipedia.org/ wiki/Stuxnet.

13. "Türk Genelkurmay Başkanlığı 27 Nisan 2007 Tarihli Basın Açıklaması," *Wikisource,* accessed June 30, 2014, accessed September 30, 2014, http://tr.wikisource.org/wiki/T%C3%BCrk_Genelkurmay_Ba%C5% 9Fkanl%C4%B1%C4%9F%C4%B1_27_Nisan_2007_tarihli_bas%C4% B1n_a%C3%A7%C4%B1klamas%C4%B1.

14. "Turkey PM Erdogan Defiant over Twitter Ban," *Al Jazeera,* March 23, 2014, accessed September 30, 2014, http://www.aljazeera.com/ news/middleeast/2014/03/turkey-pm-erdogan-defiant-over-twitter -ban-201432316413858662o.html.

15. Xeni Jardin, "Pro-Assad 'Syrian Electronic Army' Boasts of Attacks on NYT, Twitter, Huffington Post," *Boing Boing,* August 27, 2013, accessed September 30, 2014, http://boingboing.net/2013/08/27/syrian-electronic -army-boa.html; Christine Haughney and Nicole Perlroth, "Times Site Is Disrupted in Attack by Hackers," *New York Times,* August 27, 2013, accessed September 30, 2014, http://www.nytimes.com/2013/08/28/ business/media/hacking-attack-is-suspected-on-times-web-site.html.

16. Karen Deyoung and Claudia Duque, "U.S. Aid Implicated in Abuses of Power in Colombia," *Washington Post,* August 20, 2011, accessed Sep-

tember 30, 2014, http://www.washingtonpost.com/pb/national/
national-security/us-aid-implicated-in-abuses-of-power-in-colombia/
2011/06/21/gIQABrZpSJ_story.html.

17. Fruzsina Eördögh, "Israeli Defense Forces Announce Major Assault on
Gaza via Twitter, Live-Blog the Whole Thing," *Slate*, November 14, 2012,
accessed September 30, 2014, http://www.slate.com/blogs/future
_tense/2012/11/14/idf_announces_gaza_assault_death_of_ahmed
_al_jabari_via_twitter.html.

18. *Tactical Technology Collective*, accessed June 27, 2014, https://www
.tacticaltech.org/.

19. "Images Reveal Nigeria Army Abuse," *BBC News*, May 1, 2013, accessed
September 30, 2014, http://www.bbc.co.uk/news/world-africa
-22366016.

20. "Human Rights Watch Identifies Isis Execution Sites Based on Satellite Images," *Guardian*, June 27, 2014, accessed September 30, 2014,
http://www.theguardian.com/world/video/2014/jun/27/human
-rights-watch-isis-execution-sites-satellite-images-iraq-video.

21. "1/2 Drone Launched by Protesters at Warsaw, Poland," *YouTube*, November 18, 2011, accessed September 30, 2014, http://youtu.be/
03OB_4BT1LA.

22. Brittany Fiore-Silfvast, "User-Generated Warfare: A Case of Converg-
ing Wartime Information Networks and Coproductive Regulation on
YouTube," *International Journal of Communication* 6 (2012), 1965–88, accessed September 30, 2014, http://ijoc.org/index.php/ijoc/article/
view/1436.

23. *Tactical Technology Collective*.

24. FrontlineSMS: FrontlineCloud, accessed June 27, 2014, http://www.front
linesms.com/.

25. *Mobilisation Lab*, accessed June 27, 2014, http://www.mobilisationlab
.org/.

26. James C. Scott, *Seeing Like a State: How Certain Schemes to Improve the Human
Condition Have Failed* (New Haven: Yale University Press, 1998).

27. Merlyna Lim, "Clicks, Cabs, and Coffee Houses: Social Media and Oppositional Movements in Egypt, 2004–2011," *Journal of Communication* 62,

no. 2 (April 1, 2012): 231–48, doi:10.1111/j.1460-2466.2012.01628.x; Zeynep Tufekci and Christopher Wilson, "Social Media and the Decision to Participate in Political Protest: Observations From Tahrir Square," *Journal of Communication* 62, no. 2 (April 1, 2012): 363–79, doi:10.1111/j.1460-2466.2012.01629.x.

28. Catie Snow Bailard, "A Field Experiment on the Internet's Effect in an African Election: Savvier Citizens, Disaffected Voters, or Both?" *Journal of Communication* 62, no. 2 (April 1, 2012): 330–44, doi:10.1111/j.1460-2466.2012.01632.x.

29. Jonathan Hassid, "Safety Valve or Pressure Cooker? Blogs in Chinese Political Life," *Journal of Communication* 62, no. 2 (April 1, 2012): 212–30, doi:10.1111/j.1460-2466.2012.01634.x; Sebastián Valenzuela, Arturo Arriagada, and Andrés Scherman, "The Social Media Basis of Youth Protest Behavior: The Case of Chile," *Journal of Communication* 62, no. 2 (April 1, 2012): 299–314, doi:10.1111/j.1460-2466.2012.01635.x.

30. "Nigeria: Hints of a New Chapter," *Economist*, November 12, 2009, accessed September 14, 2014, http://www.economist.com/node/14843563.

31. Robert Jensen, "The Digital Provide: Information (Technology), Market Performance, and Welfare in the South Indian Fisheries Sector," *Quarterly Journal of Economics* 122, no. 3 (August 1, 2007): 879–924, doi:10.1162/qjec.122.3.879; Aparajita Goyal, *Information, Direct Access to Farmers, and Rural Market Performance in Central India* (Rochester, NY: World Bank, May 2010), accessed September 30, 2014, http://papers.ssrn.com/abstract=1613083.

32. Beth Simone Noveck, *Wiki Government: How Technology Can Make Government Better, Democracy Stronger, and Citizens More Powerful* (Washington, DC: Brookings Institution Press, 2010).

33. Mike Nizza and Patrick Lyon, "In an Iranian Image, a Missile Too Many," *Lede*, July 10, 2008, http://thelede.blogs.nytimes.com/2008/07/10/in-an-iranian-image-a-missile-too-many/.

34. Josh Chin, "Tiananmen Effect: 'Big Yellow Duck' a Banned Term," *China Real Time Report*, June 4, 2013, accessed September 30, 2014,

http://blogs.wsj.com/chinarealtime/2013/06/04/tiananmen-effect
-big-yellow-duck-a-banned-term/.

35. Ibid.

36. Jean Comaroff and John L. Comaroff, *Of Revelation and Revolution*, vol. 1,
Christianity, Colonialism, and Consciousness in South Africa (Chicago: University
of Chicago Press, 1991).

37. Manuel Castells, *Communication Power* (Oxford: Oxford University Press,
2011).

38. Christopher M. Schroeder, *Startup Rising: The Entrepreneurial Revolution Remaking the Middle East* (New York: Palgrave Macmillan, 2013).

39. Daniela Stockmann, *Media Commercialization and Authoritarian Rule in China*
(New York: Cambridge University Press, 2013).

40. Cathy Hong, "New Political Tool: Text Messaging," *Christian Science
Monitor*, June 30, 2005, accessed September 14, 2014, http://www
.csmonitor.com/2005/0630/p13s01-stct.html.

41. Cathy Hong, "New Political Tool: Text Messaging," *USA Today*, June 30,
2005, accessed September 30, 2014, http://usatoday30.usatoday.com/
tech/news/2005-06-30-politics-text-tool_x.htm.

42. Jason Gilmore and Philip N. Howard, *Does Social Media Make a Difference
in Political Campaigns? Digital Dividends in Brazil's 2010 National Elections* (Seattle: Center for Communication and Civic Engagement, June 2013),
accessed September 30, 2014, http://papers.ssrn.com/sol3/papers.
cfm?abstract_id=2273832; James Elliot, Robert O'Brien, and Jean
Stockard, eds., *Sociological Perspectives* (Thousand Oaks, CA: Sage, n.d.).

43. Fabio Rojas, "More Tweets, More Votes—It Works for Google Searches,
Too!" *Orgtheory.net*, February 14, 2014, accessed September 30, 2014,
http://orgtheory.wordpress.com/2014/02/14/more-tweets-more
-votes-it-works-for-google-searches-too/.

44. Liu Yazhou, "The Internet Has Become the Main Battlefield of Ideological Struggle," *China Scope*, no. 66 (October 15, 2013): 32; John
Garnaut, "China Must Reform or Die," *Sydney Morning Herald*, August 12,
2010, accessed September 30, 2014, http://www.smh.com.au/world/
china-must-reform-or-die-20100811-11zxd.html.

45. "A Dangerous Year," *Economist*, January 28, 2012, accessed September 30, 2014, http://www.economist.com/node/21543477.

46. Philip N. Howard and Adrienne Massanari, "Learning to Search and Searching to Learn: Income, Education, and Experience Online," *Journal of Computer-Mediated Communication* 12, no. 3 (April 1, 2007): 846–65, doi:10.1111/j.1083-6101.2007.00353.x.

47. Adrienne L. Massanari and Philip N. Howard, "Information Technologies and Omnivorous News Diets over Three U.S. Presidential Elections," *Journal of Information Technology & Politics* 8, no. 2 (2011): 177–98, doi: 10.1080/19331681.2011.541702.

48. Philip N. Howard and Steve Jones, *Society Online: The Internet in Context* (Thousand Oaks, CA: Sage, 2004).

49. Lev Muchnik, Sinan Aral, and Sean J. Taylor, "Social Influence Bias: A Randomized Experiment," *Science* 341, no. 6146 (2013): 647–51, doi:10.1126/science.1240466.

50. Nic Halverson, "Ex-Neo-Nazis, Ex-Terrorists Get Social Network," *Discovery News*, April 26, 2012, accessed September 30, 2014, http://news.discovery.com/tech/against-violent-extremism-120426.htm.

51. "Quilliam Foundation," accessed June 27, 2014, http://www.quilliamfoundation.org/.

52. Leopoldina Fotunati, Raul Pertierra, and Jane Vincent, *Migration, Diaspora, and Information Technology in Global Societies* (New York: Routledge, 2012).

53. "Stingray Tracking Devices: Who's Got Them?" *American Civil Liberties Union*, accessed June 23, 2014, https://www.aclu.org/maps/stingray-tracking-devices-whos-got-them.

54. Laura Payton, "Spy Agencies, Prime Minister's Adviser Defend Wi-Fi Data Collection," *CBC News*, February 3, 2014, accessed September 30, 2014, http://www.cbc.ca/1.2521166.

55. Google, *Google Transparency Report*, accessed September 30, 2014, http://www.google.com/transparencyreport/.

56. "A Threat to the Entire Country," *Economist*, September 29, 2012, accessed September 30, 2014, http://www.economist.com/node/21563751.

57. Heba Afify, "Khaled Said's Killing Draws 7-Year Sentences for 2 Officers," *New York Times*, October 26, 2011, accessed September 30, 2014,

http://www.nytimes.com/2011/10/27/world/middleeast/khaled
-saids-killing-draws-7-year-sentences-for-2-officers.html.

58. "Death of Neda Agha-Soltan," *Wikipedia*, accessed June 26, 2014, http://en.wikipedia.org/wiki/Death_of_Neda_Agha-Soltan.

59. Liam Stack, "Video of Tortured Boy's Corpse Deepens Anger in Syria," *New York Times*, May 30, 2011, accessed September 30, 2014, http:// www.nytimes.com/2011/05/31/world/middleeast/31syria.html.

60. Mancur Olson, *The Logic of Collective Action: Public Goods and the Theory of Groups* (Cambridge: Harvard University Press, 2009).

61. Ethan Zuckerman, *Rewire: Digital Cosmopolitans in the Age of Connection* (New York: Norton, 2013).

62. W. Lance Bennett and Alexandra Segerberg, *The Logic of Connective Action: Digital Media and the Personalization of Contentious Politics* (New York: Cambridge University Press, 2014).

63. "DeadUshahidi: Neither Dead Right nor Dead Wrong," *iRevolution*, July 5, 2012, http://irevolution.net/2012/07/05/deadushahidi/.

64. F. Edwards, Philip N. Howard, and Mary Joyce, *Digital Activism and Non-Violent Conflict* (Seattle: Digital Activism Research Project, November 2013), accessed September 30, 2014, http://digital-activism .org/2013/11/report-on-digital-activism-and-non-violent-conflict/.

65. David M. Lazer et al., "The Parable of Google Flu: Traps in Big Data Analysis," *Science* 343, no. 6176 (2014): 1203–5, doi:10.1126/science.1248506.

66. June Gates, "Federal Reserve Banks Announce New Study to Examine Nation's Payments Usage" (Washington, DC: Federal Reserve Financial Services Policy Committee, January 17, 2013), accessed September 30, 2014, http://www.federalreserve.gov/newsevents/press/ other/20130117a.htm.

67. Philip N. Howard, *New Media Campaigns and the Managed Citizen* (New York: Cambridge University Press, 2005).

68. *Premise*, accessed June 30, 2014, http://www.premise.com/.

69. Monty Munford, "M-Paisa: Ending Afghan Corruption, One Text at a Time," *TechCrunch*, October 17, 2010, accessed September 30, 2014, http:// techcrunch.com/2010/10/17/m-paisa-ending-afghan-corruption-one -text-at-a-time/.

70. "Warmed-up Numbers," *Economist*, June 23, 2012, accessed September 30, 2014, http://www.economist.com/node/21557366#.

71. Bernd Debusmann, "Betting on Syria's Assad Staying in Power," *Reuters*, February 10, 2012, accessed September 30, 2014, http://in.reuters.com/article/2012/02/10/column-debusmann-idINDEE8190FA20120210; "What Makes Heroic Strife," *Economist*, April 21, 2012, accessed September 30, 2014, http://www.economist.com/node/21553006.

5. Five Consequences of the Pax Technica

1. For more on the empires of contemporary media, see Tim Wu, *The Master Switch: The Rise and Fall of Information Empires* (New York: Vintage, 2010).

2. Boston Consulting Group, *The Internet Economy in the G-20*, 2012, accessed September 30, 2014, http://www.bcg.com/documents/file100409.pdf.

3. "IPv6," *Wikipedia*, accessed June 30, 2014, http://en.wikipedia.org/wiki/IPv6.

4. "State Enemies Archive—The Enemies of Internet," *The Enemies of the Internet: Reporters Without Borders*, 2012, accessed September 30, 2014, http://surveillance.rsf.org/en/category/state-enemies/.

5. Brian Fung, "Reporters Without Borders: For the First Time, America Is an Enemy of the Internet," *Washington Post*, March 14, 2014, accessed September 30, 2014, http://www.washingtonpost.com/blogs/the-switch/wp/2014/03/14/reporters-without-borders-for-the-first-time-america-is-an-enemy-of-the-internet/.

6. "Greatest Hits," *Economist*, May 24, 2007, accessed September 30, 2014, http://www.economist.com/node/9228794.

7. U.S. Department of State, *Saudi Arabia*, Report (Washington, DC: Department of State, Office of Website Management, Bureau of Public Affairs, June 2012), accessed September 30, 2014, http://www.state.gov/e/eb/rls/othr/ics/2012/191229.htm.

8. World Bank, "Mobile Payments Go Viral: M-Pesa in Kenya," *Yes Africa Can: Stories from a Dynamic Continent* (Washington, DC: World Bank, January 2013), accessed September 30, 2014, http://web.worldbank.org/

WBSITE/EXTERNAL/COUNTRIES/AFRICAEXT/o,,contentMDK:2255
1641~pagePK:146736~piPK:146830~theSitePK:258644,00.html.

9. "Airtime—Cellphone Banking—FNB," accessed June 30, 2014, https://
   www.fnb.co.za/ways-to-bank/airtime.html.

10. "iHub," accessed June 30, 2014, http://www.ihub.co.ke/research.

11. "Vital for the Poor," *Economist*, November 10, 2012, accessed September 30, 2014, http://www.economist.com/news/middle-east-and
   -africa/21566022-report-describes-sacrifices-poor-make-keep-mobile
   -phone-vital.

12. World Bank, "Mobile Usage at the Base of the Pyramid in Kenya," *infoDev: Growing Innovation* (Washington, DC: World Bank, December 2012), accessed September 30, 2014, http://www.infodev.org/
   infodev-files/final_kenya_bop_study_web_jan_02_2013_0.pdf.

13. "Map Kibera," accessed June 20, 2014, http://mapkibera.org/.

14. David Daw, "10 Twitter Bot Services to Simplify Your Life," *TechHive*, October 23, 2011, accessed September 30, 2014, http://www.techhive
   .com/article/242338/10_twitter_bot_services_to_simplify_your_life
   .html.

15. Rachel Haot, "Open Government Initiatives Helped New Yorkers Stay Connected During Hurricane Sandy," *TechCrunch*, January 11, 2013, accessed September 30, 2014, http://techcrunch.com/2013/01/11/
   data-and-digital-saved-lives-in-nyc-during-hurricane-sandy/.

16. Samuel P. Huntington, *The Clash of Civilizations and the Remaking of World Order* (New York: Simon and Schuster, 2011).

17. "Global Voices," accessed June 30, 2014, http://globalvoicesonline
   .org.

18. John Perry Barlows, "Declaration of Independence of Cyberspace," 1996, accessed September 30, 2014, http://wac.colostate.edu/rhetnet/
   barlow/barlow_declaration.html; Hillary Clinton, "Internet Freedom: The Prepared Text of Secretary of State Hillary Rodham Clinton's Speech, Delivered at the Newseum in Washington, DC," *Foreign Policy*, January 21, 2010, accessed September 30, 2014, http://www.foreign
   policy.com/articles/2010/01/21/internet_freedom.

19. "Aga Khan Development Network," accessed June 30, 2014, http://www.akdn.org/.

20. "Telecomix," accessed June 30, 2014, http://telecomix.org/; "Centre for Applied Nonviolent Action and Strategies," Wikipedia, accessed June 6, 2014, http://en.wikipedia.org/wiki/Centre_for_Applied _Nonviolent_Action_and_Strategies.

21. "Nawaat," accessed June 30, 2014, http://nawaat.org/portail/; Gigi Ibrahim, "Piggiepedia," Tahrir and Beyond, July 5, 2011, accessed September 30, 2014, http://theangryegyptian.wordpress.com/category/piggiepedia/.

22. Muzammil M. Hussain and Philip N. Howard, State Power 2.0 (London: Ashgate, 2013).

23. "Access: Mobilizing for Global Digital Freedom," accessed June 30, 2014, https://www.accessnow.org/.

24. "Pirate Party International," accessed June 30, 2014, http://www .pp-international.net/.

25. Ibid.

26. CBC News, "Higgins on Social Media and Syria," accessed June 30, 2014, http://podcast.cbc.ca/mp3/podcasts/current_20131216_73168 .mp3.

27. "Uchaguzi," accessed June 30, 2014, https://uchaguzi.co.ke/.

28. Steven Livingston, Africa's Information Revolution: Implications for Crime, Policing, and Citizen Security (Washington, DC: Africa Center for Strategic Studies, November 2013), accessed September 30, 2014, http://oai.dtic .mil/oai/oai?verb=getRecord&metadataPrefix=html&identifier=AD A588374.

29. "India Kanoon," accessed June 30, 2014, http://www.indiankanoon .org/about.html.

30. "LizaAlert," accessed June 27, 2014, http://lizaalert.org/.

31. "Holoda Rynda," accessed June 30, 2014, http://holoda.rynda.org/; "Rynda," accessed June 30, 2014, http://rynda.org/.

32. "Not in My Country," accessed June 30, 2014, https://www.notinmy country.org/.

33. "I Paid a Bribe," accessed June 30, 2014, http://www.ipaidabribe .com/#gsc.tab=0; "I Paid a Bribe: Kenya," accessed June 30, 2014,

http://ipaidabribe.or.ke/; "I Paid a Bribe: Pakistan," accessed June 30, 2014, http://www.ipaidbribe.pk/.

34. "Kiirti," accessed June 30, 2014, http://www.kiirti.org/static/view/whatiskiirti.

35. Zeynep Tufekci and Christopher Wilson, "Social Media and the Decision to Participate in Political Protest: Observations from Tahrir Square," *Journal of Communication* 62, no. 2 (April 1, 2012): 363–79, doi: 10.1111/j.1460-2466.2012.01629.x.

36. W. Lance Bennett, Regina G. Lawrence, and Steven Livingston, *When the Press Fails: Political Power and the News Media from Iraq to Katrina* (Chicago: University of Chicago Press, 2008).

37. "Safecast," accessed June 30, 2014, http://blog.safecast.org/.

38. John Sutter, "Google Takes on the Drug Cartels," CNN, July 19, 2012, accessed September 30, 2014, http://www.cnn.com/2012/07/19/tech/web/google-ideas-crime/index.html.

39. Alex Pentland, *Social Physics: How Good Ideas Spread—The Lessons from a New Science* (New York: Penguin, 2014).

40. danah boyd and Kate Crawford, "Critical Questions for Big Data: Provocations for a Cultural, Technological, and Scholarly Phenomenon," *Information, Communication, and Society* 15, no. 5 (2012): 662–79, doi:0.1080/1369118X.2012.678878.

41. Lee Rainie et al., *Anonymity, Privacy, and Security Online*, Internet Project (Washington, DC: Pew Research, September 2013), accessed September 30, 2014, http://www.pewinternet.org/2013/09/05/anonymity-privacy-and-security-online/.

42. Ronald J. Deibert, *Black Code: Inside the Battle for Cyberspace* (Toronto: Signal, 2013).

43. Stacey Knobler et al., *Learning from SARS: Preparing for the Next Disease Outbreak: Workshop Summary* (Washington, DC: National Academies Press, January 2004), accessed September 30, 2014, http://www.nap.edu/openbook.php?record_id=10915.

44. David Leinweber, "Stupid Data Miner Tricks: How Quants Fool Themselves and the Economic Indicator in Your Pants," *Forbes*, July 24, 2012, accessed September 30, 2014, http://www.forbes.com/sites/

davidleinweber/2012/07/24/stupid-data-miner-tricks-quants-fooling -themselves-the-economic-indicator-in-your-pants/.

45. Juan O. Tamayo, "Fiber-Optic Cable Benefiting Only Cuban Government," *Miami Herald*, May 26, 2012, accessed September 30, 2014, http://www.miamiherald.com/2012/05/25/2817534/fiber-optic -cable-benefiting-only.html.

46. Adam Tanner, "U.S.-Style Personal Data Gathering Is Spreading Worldwide," *Forbes*, October 16, 2013, accessed September 30, 2014, http:// www.forbes.com/sites/adamtanner/2013/10/16/u-s-style-personal -data-gathering-spreading-worldwide/.

47. Tom Bateman, "Police Warn over Drugs Cyber-Attack," *BBC News*, October 16, 2013, accessed September 30, 2014, http://www.bbc.co.uk/ news/world-europe-24539417.

6. Network Competition and the Challenges Ahead

1. Xiaoying Zhou, "A Hilarious Coded Riff on China's Government: 'Going Shopping for the 18th Time,'" *Tea Leaf Nation*, October 14, 2012, accessed September 30, 2014, http://www.tealeafnation.com/2012/11/a-hilarious -coded-riff-on-chinas-government-going-shopping-for-the-18th-time/.

2. Lily Kuo, "China Has More Internet Monitors Than Soldiers," *Quartz*, October 8, 2013, accessed September 30, 2014, http://qz.com/132590/ china-has-more-internet-monitors-than-active-army-personnel/.

3. Ibid.

4. Brian Kretal, Paton Adams, and George Bakos, *Occupying the Information High Ground: Chinese Capabilities for Computer Network Operations and Cyber Espionage* (Falls Church, VA: Northrop Grumman Corporation, March 2012), accessed September 30, 2014, http://origin.www.uscc.gov/sites/default/ files/Research/USCC_Report_Chinese_Capabilities_for_Computer _Network_Operations_and_Cyber_%20Espionage.pdf.

5. Graham Webster, "Five Points on the Deeply Flawed U.S. Congress Huawei Report," *Transpacifica*, October 10, 2012, accessed September 30, 2014, http://transpacifica.net/2012/10/10/five-points-on-the-deeply-flawed -u-s-congress-huawei-report/.

6. Gary King, Jennifer Pan, and Margaret E. Roberts, "How Censorship in China Allows Government Criticism but Silences Collective Expression," *American Political Science Review* 107, no. 02 (2013): 326–43, doi: 10.1017/S0003055413000014.

7. "Internet Censorship in China," Wikipedia, accessed September 30, 2014, http://en.wikipedia.org/wiki/Internet_censorship_in_China.

8. James Fallows, "On the 'Slow' Chinese Internet and the Prospects for China: One More Round," *Atlantic*, May 30, 2012, accessed September 30, 2014, http://www.theatlantic.com/international/archive/2012/05/on-the-slow-chinese-internet-and-the-prospects-for-china-one-more-round/257878/.

9. *Akamai State of the Internet Report Quarter 1: 2014* (Cambridge, MA: Akamai, January 2014), accessed September 30, 2014, http://www.akamai.com/stateoftheinternet/; Brian Spegele and Paul Mozur, "China Hardens Grip Before Meeting," *Wall Street Journal*, November 10, 2012, accessed September 30, 2014, http://online.wsj.com/news/articles/SB10001424052970204707104578092461228569642.

10. Adam Segal, "The People's Republic of Hacking," *Foreign Policy*, January 31, 2013, accessed September 30, 2014, http://www.foreignpolicy.com/articles/2013/01/31/the_people_s_republic_of_hacking_china_new_york_times?wp_login_redirect=0.

11. Douglas Farah and Andy Mosher, *Winds from the East: How the People's Republic of China Seeks to Influence the Media in Africa, Latin America, and Southeast Asia* (Washington, DC: Center for International Media Assistance, September 2010), accessed September 30, 2014, http://cima.ned.org/sites/default/files/CIMA-China-Report_1.pdf.

12. http://www.tealeafnation.com/tealeafnation.com.

13. Keith Bradsher, "China Blocks Web Access to Times," *New York Times*, October 25, 2012, accessed September 30, 2014, http://www.nytimes.com/2012/10/26/world/asia/china-blocks-web-access-to-new-york-times.html.

14. Bloomberg News, "Xi Jinping Millionaire Relations Reveal Fortunes of Elite," *Bloomberg*, June 29, 2012, accessed September 30, 2014, http://

www.bloomberg.com/news/2012-06-29/xi-jinping-millionaire
-relations-reveal-fortunes-of-elite.html.

15. William Wan, "Georgetown Students Shed Light on China's Tunnel System for Nuclear Weapons," *Washington Post*, November 29, 2009, accessed September 30, 2014, http://www.washingtonpost.com/world/national -security/georgetown-students-shed-light-on-chinas-tunnel-system-for -nuclear-weapons/2011/11/16/gIQA6AmKAO_story.html.

16. http://personaldemocracy.com/jessica-beinecke.

17. http://youtu.be/TorXQUboSXo, http://youtu.be/LPIovmLzGWo.

18. Pontus Wallin, "Crowd Sourcing or Cadre Sourcing: Why the Chinese Government Cannot Rely on the Internet for Information" (presented at the International Studies Association, San Diego, 2012).

19. http://www.interseliger.com/.

20. Philip N. Howard, "Social Media and the New Cold War," *The Great Debate*, 2012, accessed September 30, 2014, http://blogs.reuters.com/ great-debate/2012/08/01/social-media-and-the-new-cold-war/.

21. http://www.pinterest.com/pin/379569074817799499/.

22. Thomas Grove, "Russian Wikipedia Closes Site to Protest Internet Law," *Reuters*, July 10, 2012, accessed September 30, 2014, http://www .reuters.com/article/2012/07/10/net-us-russia-wikipedia-protest -idUSBRE8690NY20120710.

23. http://youtu.be/jxf-nRTDvGQ.

24. Paul Roderick Gregory, "Inside Putin's Campaign of Social Media Trolling and Faked Ukrainian Crimes," *Forbes*, May 11, 2014, accessed September 30, 2014, http://www.forbes.com/sites/paulroderickgregory/2014/05/11/ inside-putins-campaign-of-social-media-trolling-and-faked-ukrainian -crimes/.

25. Freedom House, *Venezuela*, Freedom on the Net 2013 (Washington, DC: Freedom House, January 2013), accessed September 30, 2014, http:// www.freedomhouse.org/report/freedom-net/2013/venezuela#. U7DvovmSxad.

26. Katherine Musslewhite, "Webcams Can Record Secretly," *Washington Post*, accessed June 30, 2014, accessed September 30, 2014, http://

www.washingtonpost.com/posttv/business/technology/webcams
-can-record-secretly/2013/12/18/3a48220c-6771–11e3-ae56–22deo
72140a2_video.html.

27. Dean Nelson, "China 'Hacking Websites in Hunt for Tibetan Dissidents,'" *Telegraph*, August 13, 2013, accessed September 30, 2014, http://
www.telegraph.co.uk/news/worldnews/asia/china/10240404/
China-hacking-websites-in-hunt-for-Tibetan-dissidents.html.

28. Iain Thomson, "AntiLeaks Boss: We'll Keep Pummeling WikiLeaks and Assange," *Register*, August 13, 2012, accessed September 30, 2014, http://www
.theregister.co.uk/2012/08/13/antileaks_wikileaks_attack_response/.

29. Brian Krebs, "Amnesty International Site Serving Java Exploit," *Krebs on Security*, December 22, 2011, accessed September 30, 2014, http://krebson
security.com/2011/12/amnesty-international-site-serving-java-exploit/.

30. @indiankanoon, "IK Servers Are Getting DDoSed Using the DNS Reflec-
tion Attack," *Indian Kanoon* (October 19, 2013), accessed September 30,
2014, https://twitter.com/indiankanoon/status/391497714451492865.

31. Eli Pariser, *The Filter Bubble: How the New Personalized Web Is Changing What We
Read and How We Think* (London: Penguin, 2011).

32. Keith Wagstaff, "1 in 10 Twitter Accounts Is Fake, Say Researchers,"
NBC News, November 26, 2013, accessed September 30, 2014, http://
www.nbcnews.com/technology/1-10-twitter-accounts-fake-say
-researchers-2D11655362; Won Kim et al., "On Botnets," in *Proceedings of the 12th International Conference on Information Integration and Web-Based
Applications and Services* (New York: ACM, 2010), 5–10, accessed September 30, 2014, http://dl.acm.org/citation.cfm?id=1967488.

33. Zi Chu et al., "Who Is Tweeting on Twitter: Human, Bot, or Cy-
borg?" in *Proceedings of the 26th Annual Computer Security Applications Conference*
(New York: ACM, 2010), 21–30, accessed September 30, 2014, http://
dl.acm.org/citation.cfm?id=1920265.

34. Joan Woodbrey Crocker, "What's Your Egg Count? Block Spam Bots Following You on Twitter," *Flyte New Media*, accessed September 30, 2014,
http://www.takeflyte.com/flyte/2011/08/whats-your-egg-count-block
-spam-bots-following-you-on-twitter.html.

35. Rob Dubbin, "The Rise of Twitter Bots," *New Yorker Blogs*, November 15, 2013, accessed September 30, 2014, http://www.newyorker.com/online/blogs/elements/2013/11/the-rise-of-twitter-bots.html.

36. Anas Qtiesh, "Spam Bots Flooding Twitter to Drown Info About #Syria Protests," *Anas Qtiesh's Blog*, April 18, 2011, accessed September 30, 2014, http://www.anasqtiesh.com/2011/04/spam-bots-flooding-twitter-to-drown-info-about-syria-protests/; Taylor Casti, "How to Spot a Twitter Spambot," *Mashable*, November 8, 2013, accessed September 30, 2014, http://mashable.com/2013/11/08/twitter-spambots/; Jillian C. York, "Syria's Twitter Spambots," *Guardian*, April 21, 2011, accessed September 30, 2014, http://www.theguardian.com/commentisfree/2011/apr/21/syria-twitter-spambots-pro-revolution.

37. Qtiesh, "Spam Bots Flooding Twitter to Drown Info About #Syria Protests."

38. Brian Krebs, "Twitter Bots Drown Out Anti-Kremlin Tweets," *Krebs on Security*, December 8, 2011, accessed September 30, 2014, http://krebsonsecurity.com/2011/12/twitter-bots-drown-out-anti-kremlin-tweets/; Mike Orcutt, "Twitter Mischief Plagues Mexico's Election," *MIT Technology Review*, June 21, 2012, accessed September 30, 2014, http://www.technologyreview.com/news/428286/twitter-mischief-plagues-mexicos-election/; Brian Krebs, "Twitter Bots Target Tibetan Protests," *Krebs on Security*, March 20, 2012, accessed September 30, 2014, http://krebsonsecurity.com/2012/03/twitter-bots-target-tibetan-protests/; Torin Peel, "The Coalition's Twitter Fraud and Deception," *Independent Australia*, August 26, 2013, accessed September 30, 2014, http://www.independentaustralia.net/politics/politics-display/the-coalitions-twitter-fraud-and-deception,5660; "Jasper Admits to Using Twitter Bots to Drive Election Bid," *Inside Croydon*, November 26, 2012, accessed September 30, 2014, http://insidecroydon.com/2012/11/26/jasper-admits-to-using-twitter-bots-to-drive-election-bid/; W. Oremus, "Mitt Romney's Fake Twitter Follower Problem," *Slate*, July 25, 2012, accessed September 30, 2014, http://www.slate.com/blogs/future_tense/2012/07/25/mitt_romney_fake_twitter_followers_who_s_buying_them_.html; Katy Pearce, "Cyberfuckery in Azerbaijan," *Adventures in Research*, March 10, 2013, accessed September 30, 2014, http://www.katypearce.net/cyberfuckery

-in-azerbaijan/; York, "Syria's Twitter Spambots"; Choe Sang-hun, "Prosecutors Detail Attempt to Sway South Korean Election," *New York Times*, November 21, 2013, accessed September 30, 2014, http://www .nytimes.com/2013/11/22/world/asia/prosecutors-detail-bid-to-sway -south-korean-election.html.

39. Kim et al., "On Botnets."

40. Chu et al., "Who Is Tweeting on Twitter?"

41. Philip N. Howard, "Let's Make Candidates Pledge Not to Use Bots," *Reuters Blogs—The Great Debate*, January 2, 2014, accessed September 30, 2014, http://blogs.reuters.com/great-debate/2014/01/02/lets-make -candidates-pledge-not-to-use-bots/.

42. American Association of Public Opinion Researchers, "The Problem of So-Called 'Push Polls': When Advocacy Calls Are Made Under the Guise of Research," 2007, accessed September 30, 2014, http://aapor.org/ AAPOR_Statements_on_Push_Polls1/3850.htm#.U7ErMfmSxad.

43. Julie Johnsson and Mary Schlangenstein, "Malaysian Air Said to Opt Out of Boeing Jet-Data Service," *Bloomberg*, March 12, 2014, accessed September 30, 2014, http://www.bloomberg.com/news/2014-03-12/ malaysian-air-said-to-opt-out-of-boeing-plan-to-share-jets-data.html.

44. "LG Smart TVs Logging USB Filenames and Viewing Info to LG Servers," *DoctorBeet's Blog*, November 18, 2013, accessed September 30, 2014, http://doctorbeet.blogspot.hu/2013/11/lg-smart-tvs-logging-usb -filenames-and.html.

45. "Smart TVs That Send Data Without Consent Will Be Fixed: LG," *CBC News*, October 21, 2013, accessed September 30, 2014, http://www.cbc .ca/1.2434710.

46. Ericka Chickowski, "Researchers Demo How Smart TVs Can Watch You," *Guardian*, 2014, accessed September 30, 2014, http://www.theguardian .com/media-network/partner-zone-infosecurity/smart-tv-security-risks? CMP=twt_gu.

47. Tarleton Gillespie, *Wired Shut: Copyright and the Shape of Digital Culture*, rpt. ed. (Cambridge: MIT Press, 2009).

48. Lucas Radicella, "Bolivia: Borders Will Have Biometric Security Measures," *Argentina Independent*, October 24, 2012, accessed September 30,

2014, http://www.argentinaindependent.com/currentaffairs/news
fromlatinamerica/bolivia-borders-will-have-biometric-security
-measures/.

49. Silvia Higuera, "Colombia's Supreme Court Drops 2 Charges Against Former Intelligence Director in Wire Tapping Scandal Case," *Knight Center for
Journalism in the Americas*, October 9, 2013, accessed September 30, 2014,
https://knightcenter.utexas.edu/blog/00-14561-colombias-supreme
-court-drops-2-charges-against-former-intelligence-director-wire
-tapp.

50. Human Rights Watch, *"They Know Everything We Do": Telecom and Internet
Surveillance in Ethiopia* (New York: Human Rights Watch, March 2014),
accessed September 30, 2014, http://www.hrw.org/node/123977.

51. Ellen Nakashima, "Report: Web Monitoring Devices Made by U.S. Firm
Blue Coat Detected in Iran, Sudan," *Washington Post*, July 8, 2013, accessed September 30, 2014, http://www.washingtonpost.com/world/
national-security/report-web-monitoring-devices-made-by-us-firm
-blue-coat-detected-in-iran-sudan/2013/07/08/09877ad6-e7cf-11e2
-a301-ea5a8116d211_story.html.

52. Lisa Anderson, "Demystifying the Arab Spring: Parsing the Differences
between Tunisia, Egypt, and Libya," *Foreign Affairs* 90, no. 3 (2011): 1–2.

53. Faisal Irshaid, "How Isis Is Spreading Its Message Online," *BBC News*,
June 19, 2014, accessed September 30, 2014, http://www.bbc.com/
news/world-middle-east-27912569.

54. Paisley Dodds, "Extremists Flocking to Facebook for Recruits," *Yahoo News*,
August 29, 2011, accessed September 30, 2014, http://news.yahoo.com/
extremists-flocking-facebook-recruits-151504367.html.

55. "Steganography," *Wikipedia*, accessed September 30, 2014, http://en
.wikipedia.org/wiki/Steganography.

56. Xu Wu, *Chinese Cyber Nationalism: Evolution, Characteristics, and Implications*
(Lanham, MD: Lexington, 2007).

57. John Kelly and Bruce Etling, *Mapping Iran's Online Public: Politics and Culture in
the Persian Blogosphere* (Cambridge, MA: Berkman Center for Internet and
Society, April 2008), accessed September 30, 2014, http://cyber.law
.harvard.edu/publications/2008/Mapping_Irans_Online_Public.

58. Miriam Elder, "Hacked Emails Allege Russian Youth Group Nashi Paying Bloggers," *Guardian*, February 7, 2012, accessed September 30, 2014, http://www.theguardian.com/world/2012/feb/07/hacked-emails-nashi-putin-bloggers.

59. Linda Lye, "Documents Reveal Unregulated Use of Stingrays in California," *American Civil Liberties Union*, March 13, 2014, accessed September 30, 2014, https://www.aclu.org/blog/national-security-technology-and-liberty/documents-reveal-unregulated-use-stingrays-california.

60. Christine Clarridge, "Seattle Grounds Police Drone Program," *Seattle Times*, February 7, 2013, accessed September 30, 2014, http://seattletimes.com/html/localnews/2020312864_spddronesxml.html; Lye, "Documents Reveal Unregulated Use of Stingrays in California."

7. Building a Democracy of Our Own Devices

1. Marcus Wohlsen, "Why Copyrighted Coffee May Cripple the Internet of Things," *Wired*, March 6, 2014, accessed September 30, 2014, http://www.wired.com/2014/03/copyrighted-coffee-undermine-whole-internet-things/.

2. Tarleton Gillespie, *Wired Shut: Copyright and the Shape of Digital Culture*, rpt. ed. (Cambridge: MIT Press, 2009).

3. Robert D. Putnam, Robert Leonardi, and Raffaella Y. Nanetti, *Making Democracy Work: Civic Traditions in Modern Italy* (Princeton: Princeton University Press, 1994).

4. Steve Schifferes, "How Bretton Woods Reshaped the World," BBC, November 14, 2008, accessed September 30, 2014, http://news.bbc.co.uk/2/hi/7725157.stm.

5. "Edward Gibbon," *Wikipedia*, accessed June 27, 2014, accessed September 30, 2014, http://en.wikipedia.org/wiki/Edward_Gibbon; "The History of the Decline and Fall of the Roman Empire," *Wikipedia*, accessed June 21, 2014, http://en.wikipedia.org/wiki/The_History_of_the_Decline_and_Fall_of_the_Roman_Empire.

6. OECD, *Ensuring Fragile States Are Not Left Behind: 2013 Factsheet on Resource Flows and Trends* (Paris: OECD, January 2013), accessed September 30, 2014, http://www.oecd.org/dac/incaf/factsheet%202013%20resource%20flows%20final.pdf.

7. David Unger, "Your Thermostat is About to Get Smart," *Christian Science Monitor*, November 5, 2013, accessed September 30, 2014, http://www.csmonitor.com/Environment/Energy-Voices/2013/1105/Your-thermostat-is-about-to-get-smart-video video.

8. Rebecca MacKinnon, *Consent of the Networked: The Worldwide Struggle for Internet Freedom* (New York: Basic, 2013).

9. Rebecca MacKinnon, "Keynote Speech on Surveillance," in *Opening Ceremony of the Freedom Online Conference*, 2013, accessed September 30, 2014, http://consentofthenetworked.com/2013/06/17/freedom-online-keynote/.

10. "Aaron Swartz," *Wikipedia*, accessed June 29, 2014, http://en.wikipedia.org/wiki/Aaron_Swartz.

11. "Russian Business Network," *Wikipedia*, accessed June 19, 2014, http://en.wikipedia.org/wiki/Russian_Business_Network.

12. "Zero-Day Attack," *Wikipedia*, accessed June 21, 2014, http://en.wikipedia.org/wiki/Zero-day_attack.

13. "U.S.-Style Personal Data Gathering Is Spreading Worldwide," *Forbes*, accessed June 29, 2014, http://www.forbes.com/sites/adamtanner/2013/10/16/u-s-style-personal-data-gathering-spreading-worldwide/; Paul Schwartz, *Managing Global Privacy* (Berkeley: ThePrivacyProjects.org, January 2009), accessed September 30, 2014, http://theprivacyprojects.org/wp-content/uploads/2009/08/The-Privacy-Projects-Paul-Schwartz-Global-Data-Flows-20093.pdf.

14. "Computer Says No," *Economist*, June 22, 2013, accessed September 30, 2014, http://www.economist.com/news/international/21579816-denial-service-attacks-over-internet-are-growing-easier-and-more-powerful-their.

15. Joe Davidson, "Too Many People with Security Clearances, but Cuts Could Help Some Feds, Hurt Others," *Washington Post*, March 20, 2014, accessed September 30, 2014, http://www.washingtonpost.com/politics/federal_government/too-many-people-with-security-clearances-but-cuts-could-help-some-feds-hurt-others/2014/03/20/1f1d011a-b05e-11e3-a49e-76adc9210f19_story.html.

16. Michael Scherer, "The Geeks Who Leak," *Time*, June 24, 2013, accessed September 30, 2014, http://content.time.com/time/magazine/article/0,9171,2145506,00.html.

17. "Backdoor Dealings," *Economist*, September 14, 2013, accessed September 30, 2014, http://www.economist.com/news/leaders/21586345 -covertly-weakening-security-entire-internet-make-snooping-easier -bad.

18. Michael Riley, "NSA Said to Exploit Heartbleed Bug for Intelligence for Years," *Bloomberg*, April 12, 2014, accessed September 30, 2014, http:// www.bloomberg.com/news/2014-04-11/nsa-said-to-have-used -heartbleed-bug-exposing-consumers.html.

19. US Institute of Peace, *Non-Violent Struggle: 50 Crucial Points* (Washington, DC: United States Institute of Peace, January 2006), accessed September 30, 2014, http://www.usip.org/publications/non-violent-struggle -50-crucial-points.

20. https://www.tacticaltech.org/, https://citizenlab.org/.

21. Alan Rusbridger, "Glenn Greenwald's Partner Detained at Heathrow Airport for Nine Hours," *Guardian*, August 19, 2013, http://www.the guardian.com/world/2013/aug/18/glenn-greenwald-guardian -partner-detained-heathrow.

22. "American Middle Class," *Wikipedia*, accessed June 28, 2014, http:// en.wikipedia.org/wiki/Middle_class.

23. Alex Pentland, *Social Physics: How Good Ideas Spread: The Lessons from a New Science* (New York: Penguin, 2014).

24. Philip N. Howard, *New Media Campaigns and the Managed Citizen* (New York: Cambridge University Press, 2005).

25. Brett M. Frischmann, *Infrastructure: The Social Value of Shared Resources* (New York: Oxford University Press, 2012).

26. Richard Kielbowicz, *News in the Mail: The Press, Post Office, and Public Information, 1700–1860s* (New York: Praeger, 1989).

27. Julian Dibbell, "Internet Freedom Fighters Build a Shadow Web," *Scientific American* 306, no. 3 (March 2012).

28. "Light and Shady," *Economist*, April 21, 2012, accessed September 30, 2014, http://www.economist.com/node/21553012.

29. http://www.globalwitness.org/campaigns/corruption/anonymous -companies.

30. Kris Erickson and Philip N. Howard, "A Case of Mistaken Identity? News Accounts of Hacker, Consumer, and Organizational Responsibility for

Compromised Digital Records," *Journal of Computer-Mediated Communication* 12, no. 4 (2007): 1229–47.

31. http://tosdr.org/.

32. https://www.globalnetworkinitiative.org/.

33. "Carlos Slim," *Wikipedia*, accessed June 28, 2014, http://en.wikipedia.org/wiki/Carlos_Slim; "Megahurts," *Economist*, February 11, 2012, accessed September 30, 2014, http://www.economist.com/node/21547280.

34. "Light and Shady."

35. http://www.fcc.gov/encyclopedia/universal-service.

36. "Www.africa.slow," *Economist*, August 27, 2011, http://www.economist.com/node/21526937.

37. "Last Mile," *Wikipedia*, accessed June 19, 2014, http://en.wikipedia.org/wiki/Last_mile.

38. T. C. Sottek, "Google Now Offers Google Earth, Picasa, and Chrome in Syria," *Verge*, May 24, 2012, accessed September 30, 2014, http://www.theverge.com/2012/5/24/3041459/google-earth-picasa-chrome-syria.

39. Monk School of Global Affairs, *Internet Filtering in a Failed State: The Case of Netsweeper in Somalia* (Toronto: University of Toronto, February 2014), accessed September 30, 2014, https://citizenlab.org/2014/02/internet-filtering-failed-state-case-netsweeper-somalia/; Monk School of Global Affairs, *O Pakistan, We Stand on Guard for Thee: An Analysis of Canada-Based Netsweeper's Role in Pakistan's Censorship Regime* (Toronto: University of Toronto, June 2014), accessed September 30, 2014, https://citizenlab.org/2013/06/o-pakistan/.

40. Kate Crawford et al., "Seven Principles for Big Data and Resilience Projects," *iRevolution*, September 23, 2013, accessed September 30, 2014, http://irevolution.net/2013/09/23/principles-for-big-data-and-resilience/.

41. A PGP key allows you to securely send and receive email. "Pretty Good Privacy," *Wikipedia*, accessed June 26, 2014, http://en.wikipedia.org/wiki/Pretty_Good_Privacy.

GLOSSARY

Affordances what you can do with technology, even if designers don't anticipate that activity. Many engineers and computer scientists imagine the possible benefits of the internet of things, but a world full of device networks will also afford corporations and governments access to immense amounts of private data.

Big data information about many people collected over many kinds of devices, information that tends to reveal behavior but not attitudes or aspirations

Crowd sourcing allowing many people over many devices to provide individual pieces of information that in aggregated form have immense utility

Crypto clan an extended group of friends and family that we actively and purposefully maintain through norms of trust and reciprocity in encryption, content sharing, social networking, and information filtering

Cyberattack the process of finding and exploiting vulnerable device networks by entering them and copying, exporting, or changing the data within them

Cyberwar conflict involving the professional staff of established militaries who no longer just act in response to offline events but are trained to respond to the last cyberattack

Democracy   a form of open society in which people in authority use the internet of things for public goods and human security in ways that have been widely reviewed and publicly approved. Democracy occurs when the rules and norms of mass surveillance have been developed openly, and state practices are acknowledged by the government.

Device tithe   reserving 10 percent of processing power, sensor time, bandwidth, or other network-device feature for the user to voluntarily and openly assign to the civic organizations of his or her choice

Digital activism   an organized public effort with clear grievances, targeted authority figures, and campaigns initiated using device networks

Digital club   a small group of people with strong, direct ties who use their device networks for providing collective goods

Digital dilemma   the difficult choice of internet policy: either encourage internet access and device networking for economic benefits and political risks, or avoid civic engagement over the internet and pass up economic benefits

Dirty network   a social and technical system for corruption, crime, and human rights violations that supports negative social capital, passes negative memes, or allows for negative feedback

Ideology   meaning in the service of power. Today, information technology is the most important tool for servicing power.

Institutions   norms, rules, and patterns of behavior

Internet exchange point   the physical infrastructure that serves as a mandatory point of passage for data flowing from one network, or service provider, to others

Internet of things networks of manufactured goods with embedded power supplies, small sensors, and an address on the internet. Most of these networked devices are everyday items that are sending and receiving data about their conditions and our behavior.

Metadata information about our use of device networks. It is data about data, revealing context clues about the who, what, when, where, why, and how of the production and consumption of digital content and devices.

Net neutrality the idea that all data, devices, and people on the internet should be treated equally

Organization a social unit made up of people and material resources, like desks and device networks

Pax technica a political, economic, and cultural arrangement of institutions and networked devices in which government and industry are tightly bound in mutual defense pacts, design collaborations, standards setting, and data mining

Sociotechnical system an organization defined by the relationships between people and devices

TOR Server a free and open-source software that allows your device to anonymously connect to a network so that activities and location cannot be discovered by someone surveilling the communication

ACKNOWLEDGMENTS

The people you meet through teaching, both colleagues and students, are invaluable sources of insight. All have helped to gather stories, talked through the interpretation of data, and offered insightful comments. While working at the University of Washington, Columbia University, Princeton University, and Central European University, I have been researching and testing a cluster of propositions under the pax technica rubric with an extended team of researchers. Academic colleagues, democracy advocates, and state censors have been invaluable in sharing stories, doing interviews, and running big data analysis. Roberto Juárez-Garza wrote an excellent master's degree thesis on the use of social media in Monterrey, Mexico. Mary Joyce built an impressive database of digital-activism cases. Muzammil Hussain traveled North Africa and the Middle East for additional interviews during the Arab Spring. Marwa Mazaid proffered stories from her fieldwork in Cairo and Istanbul. Aiden Duffy and Deen Freelon crunched the "big data." Conversations with Eman Abdelrahman, Gregory Asmolov, Primož Kovačič, Patrick Meier, and Oscar Morales were inspiring. I am also grateful to democracy advocates in Azerbaijan, China, Russia, Singapore, Tunisia, Turkey, and Venezuela who were willing to meet and talk. Matthew Adeiza, Will Mari, Elisa Mason, Ruchika Tulshyan, and Sam Woolley gave great editing and research advice.

Portions of this manuscript have appeared in Philip N. Howard and Muzammil Hussain, *State Power 2.0: Digital Networks and Authoritarian Rule*

(London: Ashgate, 2013), Philip N. Howard and Muzammil Hussain, "What Best Explains Successful Protest Cascades? ICTs and the Fuzzy Causes of the Arab Spring," *International Studies Review* 15, no. 1 (2013): 48–66, and some public writing.

I am very grateful for input from my agent, Will Lippincott, and my editor, Joe Calamia. At Yale University Press, Dan Heaton edited the manuscript and taught me what grammatical expletives were (and that I had too many). Samantha Ostrowski prepared the book for release, and Nancy Ovedovitz designed the book's jacket. Their support and friendly critiques have made this a stronger manuscript. Princeton University, the University of Washington, and Central European University in Budapest provided support for my time to write, access to first-class research materials, and interaction with diverse intellectual and policy communities.

I have benefited from many different kinds of public research support for this work. This material is based upon work supported by the National Science Foundation under grant number 1144286, "RAPID—Social Computing and Political Transition in Tunisia," grant number 0713074, "Human Centered Computing: Information Access, Field Innovation, and Mobile Phone Technologies in Developing Countries," and grant number BIGDATA-1450193, "EAGER CNS—Computational Propaganda and the Production/Detection of Bots." It is based upon work supported by the U.S. Institutes of Peace under grant number 212-11F, "Digital Media, Civic Engagement, and Non-Violent Conflict." Any opinions, findings, and conclusions or recommendations expressed in this material are those of the author and do not necessarily reflect the views of the National Science Foundation or U.S. Institute of Peace.

For their feedback on earlier drafts and presentations of this work, I thank Lance Bennett, Larry Diamond, Steve Schultze, and Annemarie Slaughter. For conversations on radios, sensors, applications, and chip design I thank Josephine Bolotski, Andrew Donovan, Steven Dossick,

ACKNOWLEDGMENTS

and Dan Kasha. For their support in creating good environments for writing and research, I thank David Domke, Ed Felten, Judy Howard, Ellen Hume, and John Shattuck.

Hammer and Gordon Howard are still the key reasons for working hard (please help I am being held %#$@! . . . . .).

Budapest, Hungary

INDEX

Abdelrahman, Eman, 75–77, 78, 100, 239
Abdullah (Saudi king), 92
aboriginal communities, building collective identity, 145
AccessNow, 165
affordances, 44, 219, 240, 295
Afghanistan: anarchy in, 94; corruption in, 142–43; heroin trade in, 96
Africa: banking sector in, 102–3, 159–60; wiring of, 251–52
Aga Khan Foundation, 164
Agha-Soltan, Neda, 136–37
Aidra, 4
Airtime, 103, 159
Algeria, 92, 120–21
Al-Khateeb, Hamza Ali, 137
al Saud, Nayef, 91–92
al Saud, Salman, 92
al-Sheikh, Ali Jawad, 137
altruism, social media and, 17–18, 21–22, 195, 240
American Association of Public Opinion Researchers, 210
American Empire, 231
Amnesty International, 169, 201

anarchy, 94
Anderson, Lisa, 216
Anonymous (online activists), 43, 44–45, 117, 164
Ansar Eddine, 80
anti-cnn.com, 218
Anti-Counterfeiting Trade Agreement, 164–65
antiestablishment movements, xx
Apache, 64
Apple, 8, 63–64
April 6 Movement, 238
Arabs, cyberwarfare by, 154–55
Arab Spring, xx, 20, 28, 38, 45, 53, 60, 221, 229; Algeria's response to, 121; arrival in Egypt, 79; attracting new protestors, 174–75; born digital, 38, 51; coalescing of, 131; first protestors in, 172; governments' responses to, 155–56; organization of, 52; organizers of, 52; regulation and, 58. See also Egypt; Tunisia
Aramco, 40
Arriagada, Arturo, 122
Assad, Bashar al-, 28, 82, 116
Assange, Julian, 43–44, 238

phone companies, privatization of,
56–57, 74
Piggipedia, 164
piracy, combating, 98
Pirate Bay, 13, 166
Pirate Parties, xx, 166
pirates, 72, 81, 93–94
police, technologies leaking to, 222
political bots, 31, 204–11, 233, 234
political communication, 13
political data, xxi
political groups, digital infrastructure and, 135
political information, consumption
of, 45
political internet, xiii, 36
political order, 53, 108–10: and relationships between devices, 34;
rethinking, 224–25
political power, 16, 233
politics: criminal organizations dis-
rupting, 80; digital engagement
with, 9–10; face-to-face contact
and, 44; global, 44, 62, 149
Polity IV Project, 92
poor people, global distribution of,
97
Popović, Srđa, 238
Popular Party (Spain), 128
Premise, 142
PRI party (Mexico), 51
privacy groups, 163
privacy violations, 179–80
process tracing, 110
propaganda: bots used for, 205;
computational, 29–30; digital
media weakening, 124; geotagged, 114–15

prostitution, digital media and, 103
public alert systems, 18–19
public good, 250–51
public opinion, manipulation of, 43
public spectrum, public auctions for,
250
push polling, 209–10, 211
Pussy Riot, 86, 171
Putin, Vladimir, 171, 197, 201

Qtiesh, Anas, 204

Radonski, Henrique, 93
Rassd News Network, 20
reality-based research, 179
Recollections (Tocqueville), 108
Reporters Without Borders, 163
revolutions: histories of, 61; nature
of, 108–9
Rheingold, Howard, 85
robocalls, 207
Roh Moo-hyun, 127–28
Rolls-Royce, 212
Roman Empire, 1, 67, 107–8, 146,
231. See also Pax Romana
Romania, hacking by, 41
Romney, Mitt, 205
Rose Revolution, 238
Russia: bots in, 205; broadcast media
in, 198–99; civil society in, 171,
198; counterpropaganda in, 200;
developing network infrastructure, 183; election fraud in, 134;
election monitoring in, 199–200;
fighting piracy, 98; generating
pro-government messages on
social media, 31, 196, 200; hacking by, 41; internet in, 171, 198,